MATERIAL SCIENCE AND PROCESSES

R.K. Purohit

B.E. (Mech.), M.E.(Mech.)
M.ASME, ASCE, ASAE, IASH, LM.TSI
Department of Mechanical Engineering
M.B.M. Engineering College
JNV University
Jodhpur - 342 001 (Raj.)

SCIENTIFIC PUBLISHERS (INDIA)

P.O. BOX 91 JODHPUR

Published by:
Pawan Kumar
Scientific Publishers (India)
5-A, New Pali Road, P.O. Box 91
JODHPUR - 342 001
www.scientificpub.com

Print, 2012

ISBN: 978-81-7233-782-7

Lasertype set: Rajesh Ojha
Printed in India

In the everlasting memory of my father Prof. B.K. Purohit

PREFACE

Material Science and Processes is a core subject having close relations with all branches of Engineering. No branch of Engineering can be studied properly if one does not have indepth knowledge of material science. Based on this premise AICTE has prescribed this subject for all branches of Engineering.

The concepts of material science are based on different and diverse phenomenon related to materials and alloys with vast application in all branches of Engineering. Hence there is strong rationale for teaching this subject as an integrated topic in Engineering and relevance of materials in 'Engineering Application'.

Considering the fundamental importance of material science and processes, I have ventured to introduce this subject in simple language which can be easily digested by the beginners. The contents of the book has relevance with the subject prescribed by JNVU, Rajasthan University and Institution of Engineers as well as to the courses of study prescribed by various universities of India.

My Publisher Shri Pawan Kumar deserves my thanks for making this publication possible in a record time.

My indebtness is due to my wife Manju Purohit, son Abhishek Purohit (Software Engineer, Australia) and daughter Khushboo Purohit (Computer Science Engineer) for their co-operation and valuable suggestions.

Arrival of my grandson, Chinmay Purohit, has given me fresh vigour to embark upon the book writing project. Hence, entire credit for this book goes to the little one who, I hope, will continue to inspire me to write many more books to enrich the quality of text books on Mechanical Engineering.

I hope, the teachers and the student community would find this book useful.

JODHPUR **R.K. Purohit**

CONTENTS

Preface — *v*

1. INTRODUCTION — 1

1.1 Introduction — 1
1.2 Selecting Materials — 2
1.3 Properties of Materials — 2
1.4 Mechanical Behaviour of Materials — 3
1.5 The Tensile Test — 4
1.5.1 Yield point — 5
1.5.2 Ultimate tensile strength — 6
1.5.3 Necking — 7
1.5.4 Proportional limit — 8
1.5.5 Elastic limit — 8
1.5.6 Proof stress — 8
1.5.7 Young's modulus, E or modulus of elasticity — 9
1.6 Ductility and brittleness — 9
1.7 Toughness — 10
1.8 Hardness — 10
EXERCISES — 11

2. MECHANISM OF DEFORMATION IN CRYSTALLINE MATERIALS — 12

2.1 Introduction — 12
2.2 Crystallization — 12
2.3 Crystal Modes — 14
2.4 Atomic Radius — 18
2.5 Density of Crystal — 19
2.6 Characteristics of Crystal Structures — 19
2.6.1 Coordination number — 20
2.6.2 Atomic Packing factor — 20
2.7 Crystal Directions and Planes — 22
2.7.1 Miller Indices — 22
2.7.2 Important features of Miller indices of crystal planes — 24

2.8 The Deformation of Materials 24
2.8.1 Elastic deformation 25
2.8.2 Plastic deformation 25
2.8.3 Deformation by slip 26
2.8.4 Twinning 27
2.8.5 Difference between slip and twinning 30
2.9 Defects in Solid Crystals 30
2.9.1 Point defects 31
2.9.2 Line defects 31
2.9.3 Interfacial, plane or sheet defects 36
2.9.4 Bulk defects 36
2.10 Work Hardening (Strain Aging or Strain Hardening) 37
2.11 Grains and Grain Boundaries 38
2.12 Recovery, Recrystallization, and Grain Growth 39
2.13 Homologous Temperature Ranges for Various Process 41
2.14 Fracture 42
2.14.1 Effects of inclusions 44
2.14.2 Transition temperature 44
2.14.3 Strain aging 45
2.14.4 Defects 46
2.15 Fatigue 46
2.15.1 Fatigue failure with a metal 47
2.15.2 Stages in fatigue failure 47
2.15.3 S-N curves 48
2.15.4 Factors affecting the fatigue properties of metals 50
2.16 Creep 52
2.16.1 Factors affecting creep with metals 53
2.17 Hardness 55
2.18 Impact 57
EXERCISES 58

3. **ENGINEERING MATERIALS** 61

3.1 Introduction 61
3.2 Metals and Non-metals 61
3.3 Metals 61
3.3.1 Pure Metals 61
3.3.2 Alloys 62
3.3.3 Coated metals 62

3.3.4 Clad metals 62
3.3.5 Ferrous metals 62
3.3.6 Sintered metals 63
3.4 Phase and Structures in Steels 63
3.4.1 Ferrite 63
3.4.2 Austenite 63
3.4.3 Cementite 63
3.4.4 Eutectic (Ledeburite) 64
3.4.5 Pearlite 65
3.4.6 Martensite 65
3.4.7 δ-iron 65
3.4.8 Bainite 65
3.4.9 Sorbite and Troostite 66
3.5 Heating of Pure Iron 66
3.6 Slow cooling of Iron at different percentage of Carbon 66
3.7 Plain Carbon Steels 69
3.7.1 The eutectoid region 70
3.7.2 Time-Temperature-Transformation diagrams (T T T diagrams) 71
3.7.3 Classification of plain carbon steel and their applications and uses 73
3.7.4 Limitations of plain carbon steels 74
3.8 Alloy Steels 74
3.8.1 Low alloy steel 74
3.8.2 High alloy steels 77
3.9 Cast Irons 80
3.9.1 Types of Cast iron 80
3.9.2 Uses of Various cast irons 82
3.10 Heat Treatment of Steels 83
3.10.1 Reasons for heat treatments 84
3.10.2 The hardening steel - The martensite reaction hardening 84
3.11 Heat Treatment Processes 88
3.11.1 Annealing 88
3.11.2 Normalizing 91
3.11.3 Quenching 91
3.11.4 Tempering 92
3.11.5 Surface hardening 93
3.11.6 Case hardening 93
3.11.7 Nitriding 94
3.11.8 Flame hardening and induction hardening 94
3.11.9 Age hardening or precipitation hardening 95

3.11.10 Austempering and marquenching 95
3.12 Heat Treatment for Alloy Steel 96
3.13. Non-ferrous alloys 98
3.13.1 Advantages of non-ferrous alloys over ferrous alloys 98
3.13.2 Advantages of ferrous alloys over non-ferrous alloys 98
3.13.3 The range of non-ferrous alloys 98
3.14 Aluminium and its Alloys 98
3.14.1 Non heat treatment wrought alloys 99
3.14.2 Heat treatable wrought alloys 99
3.14.3 Age hardening in aluminium alloys 100
3.14.4 Non heat treatable cast alloys 101
3.14.5 Heat treatable cast alloys 103
3.15 Copper and its alloys 103
3.15.1 Uses of copper and its alloys 104
3.15.2 Copper alloys 104
3.15.3 Applications of copper alloys 108
3.16 Bearing Materials 109
3.16.1 Types of materials 109
EXERCISES 110

4. MANUFACTURING PROCESSES 115

4.1 Introduction 115
4.2 Forming Processes 116
4.3 Hot Working 116
4.3.1 Advantages and disadvantages of hot working 116
4.4 Hot working methods 117
4.4.1 Rolling 117
4.4.2 Forging 118
4.5 Forging Tools 121
4.6 Hand Tools 123
4.6.1 Chisels 123
4.6.2 Fullers 124
4.6.3 Swages 124
4.6.4 Flatters 125
4.6.5 Punches and drifts 126
4.6.6 Tongs 127
4.7 Forging Operations 127
4.7.1 Upsetting 127

4.7.2 Drawing down 129
4.7.3 Setting down 129
4.7.4 Bending 129
4.7.5 Piercing 131
4.7.6 Drawing 133
4.7.7 Spinning 134
4.7.8 Extrusion 134
4.8 Cold Working 139
4.8.1 Principles of cold working 139
4.8.2 Advantages and Limitations of cold working 139
4.9 Cold Working Methods 140
4.9.1 Rolling 141
4.9.2 Cold forging 141
4.9.3 Drawing (Rod, wire and tube drawing) 142
4.9.4 Bending 143
4.9.5 Spinning 144
4.9.6 Extruding 144
4.9.7 Squeezing 145
4.9.8 Peening 146
4.9.9 Sizing, coining and hobbing 146
4.10 Formability of Sheet Metals 146
4.11 Forgeability 147
4.12 Metallurgical advantages of Hot working over cold working processes 147
EXERCISES 148

5. FOUNDRY **151**

5.1 Introduction 151
5.2 Mould making 153
5.2.1 Foundry tools and equipment 153
5.2.2 The process of mould making 155
5.2.3 Moulding materials 156
5.3 Patterns 158
5.3.1 Difference between pattern and casting 159
5.3.2 Functions of a pattern 159
5.3.3 Pattern materials 159
5.3.4 Pattern allowances 163
5.3.5 Pattern types 163
5.4 Sand Casting 168

5.4.1 Moulding a bracket 168
5.4.2 Core and Core prints 170
5.4.3 Split patterns 171
5.4.4 Plate moulding 172
5.4.5 Cooling characteristics of Castings 172
EXERCISES 173

6. WELDING, BRAZING AND SOLDERING 175

6.1 Introduction 175
6.2. Welding 175
6.3 Gas Welding 176
6.4 Oxy-Acetylene Gas Welding 176
6.4.1 Gas Cylinders 177
6.4.2 Gas flames 177
6.4.3 Welding torch 180
6.4.4 Oxyacetylene welding technique 181
6.5 Arc-welding 181
6.5.1 Shielded metal arc welding (SMAW) 182
6.5.2 Comparison between A.C. and D.C. arc welding 184
6.6 Metallurgy of Welding 184
6.7 Electrodes 186
6.8 Advantages and Limitations of the Oxy-acetylene welding processes 186
6.9. Advantages and Limitations of the Manual Metallic Arc Welding 187
6.10 Brazing and Soldering 188
6.11 Brazing 188
6.11.1 Brazing metals 189
6.11.2 Filler metals 191
6.11.3 Fluxes 191
6.11.4 Brazing methods 192
6.11.5 Braze welding 193
6.11.6 Comparison with welding 193
6.11.7. Advantages of Brazing 193
6.11.8 Disadvantages of Brazing 193
6.12 Soldering 194
EXERCISES 196

INDEX 200

1

INTRODUCTION

1.1 INTRODUCTION

The history of man has been linked with his ability to work with materials, beginning with the Stone Age and ranging through the eras of copper and bronze, the Iron Age, and recently the age of steel, with our sophisticated ferrous and nonferrous materials. We are now entering the age of tailor-made materials such as composites, as indicated in Fig. 1.1. The alloys used in the manufacture of compressor airfoils since 1945 have continuously changed. As materials become more difficult to manufacture with existing manufacturing methods.

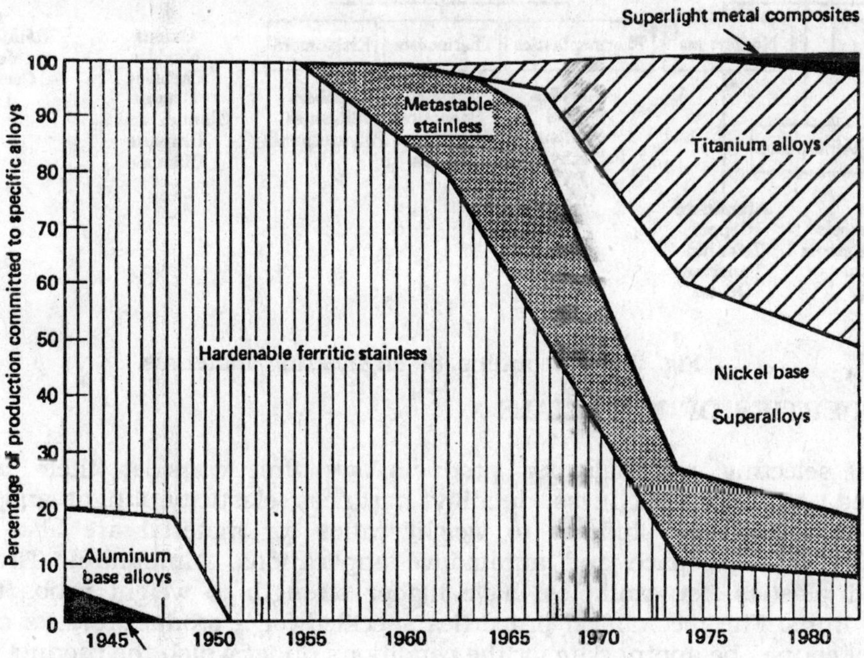

Fig. 1.1 Changing trends

1.2 SELECTING MATERIALS

An ever increasing variety of materials is now available, each having its own characteristics, applications, advantages, and limitations (Fig. 1.2). The following are the general types of materials used in manufacturing today, either individually or in combination.

1. Ferrous metals : Carbon, alloy, stainless, and tool and die steels.

2. Nonferrous metals and alloys : Aluminium, Magnesium, Copper, Nickel, Titanium, Superalloys, refractory metals, Beryllium, Zirconium, low melting alloys, and precious metals.

3. Plastics : Thermoplastics, thermosets, and elastomers.

4. Ceramics : Glass ceramics, glasses, graphite, and diamond.

5. Composite materials : Reinforced plastics, metal-matrix and ceramic matrix composites, and honey comb structures. These are also known as engineered materials.

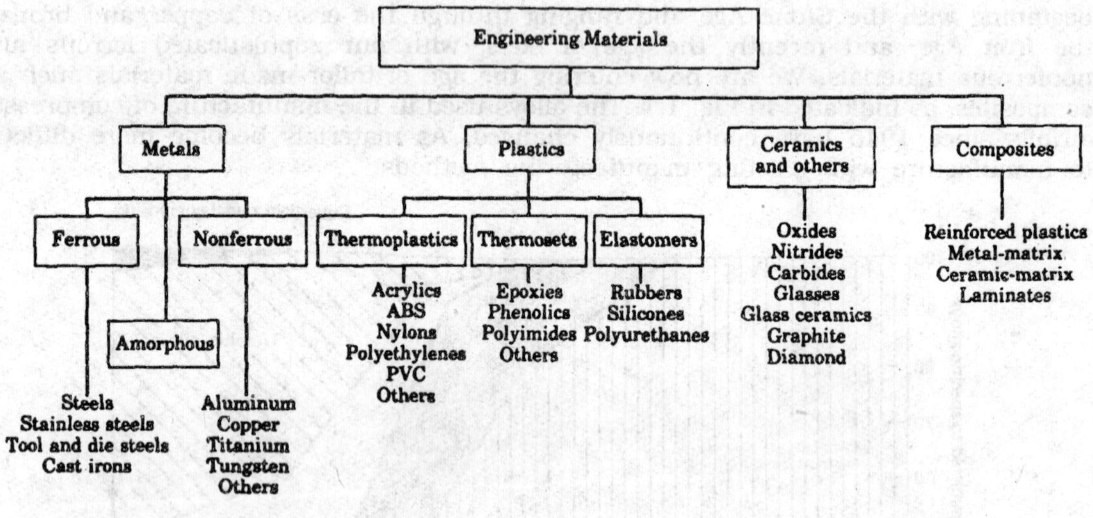

Fig. 1.2 An outline of engineering materials.

1.3 PROPERTIES OF MATERIALS

When selecting materials for products, we first consider their **mechanical properties** : strength, toughness, ductility, hardness, elasticity, fatigue and creep. The strength-to-weight and stiffness to weight ratios for material are also important, particularly for aerospace and automotive applications. Aluminium, Titanium and reinforced plastics, for examples, have higher strength to weight ratios than steels and cast irons. The mechanical properties specified for a product and its components should, of course, be appropriate for the conditions under which the product is expected to function.

We then consider the **physical properties** of density, specific heat, thermal expansion and conductivity, melting point, and electrical and magnetic properties.

Chemical properties also play a significant role in hostile as well as normal environments. Oxidation, corrosion, general degradation of properties, toxicity and flammability of materials are among the important factors to be considered.

Manufacturing properties of materials determine wheather they can be cast, formed, shaped, machined, welded, or heat treated with relative ease. Another important consideration is that the methods used to process materials to the desired shapes may adversely affect the product's final properties, service life and cost.

1.4 MECHANICAL BEHAVIOUR OF MATERIALS

In manufacturing, one of the most important groups of processes is **plastic deformations,** namely, shaping materials, by applying forces in various ways. It includes bulk deformation (forging, rolling, extrusion, and rod and wire drawing) and sheet forming processes bending, drawing, spinning and general pressworking).

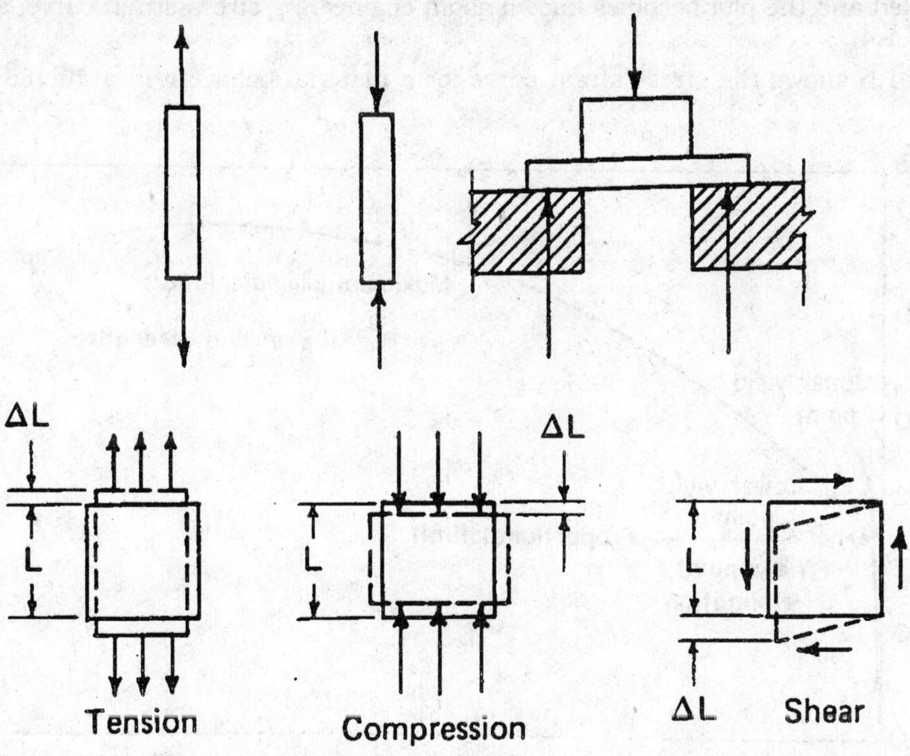

Fig. 1.3 Examples of tension, compression and shear loading, and their strain response.

Mechanical behaviour of materials during deformation deals the fundamental aspects of deformation modes, stresses, forces, effect of rate of deformation and temperature, hardness, residual stresses, and yield criteria.

In all these processes, the material is subjected to one or more of the basic modes of deformation shown in Fig. 1.3, namely, tension, compression, and shear. The degree of deformation to which the material is subjected is defined as **strain**.

1.5 THE TENSILE TEST

Considerable information about the properties of a material can be obtained from a uniaxial tensile test.

A load is applied and measured by the testing machine, while elongation or strain over the specified gauge length is determined by an external measuring device attached to the specimen. The result is a plot of coordinated load-elongation points, producing a curve of the form in Fig. 1.4. Since characteristic loads differ with different size specimens and elongations vary with different gauge lengths, it becomes desirable to remove the size effect and establish a plot, that is characteristic of the material's response to the test conditions. If the load is divided by the original cross sectional area and the elongation is divided by the original gauge length, the size effects are eliminated and the plot becomes known as an engineering stress-strain curve, as shown in Fig. 1.4.

Fig. 1.5 shows the stress strain curve for a material behaving in a ductile manner.

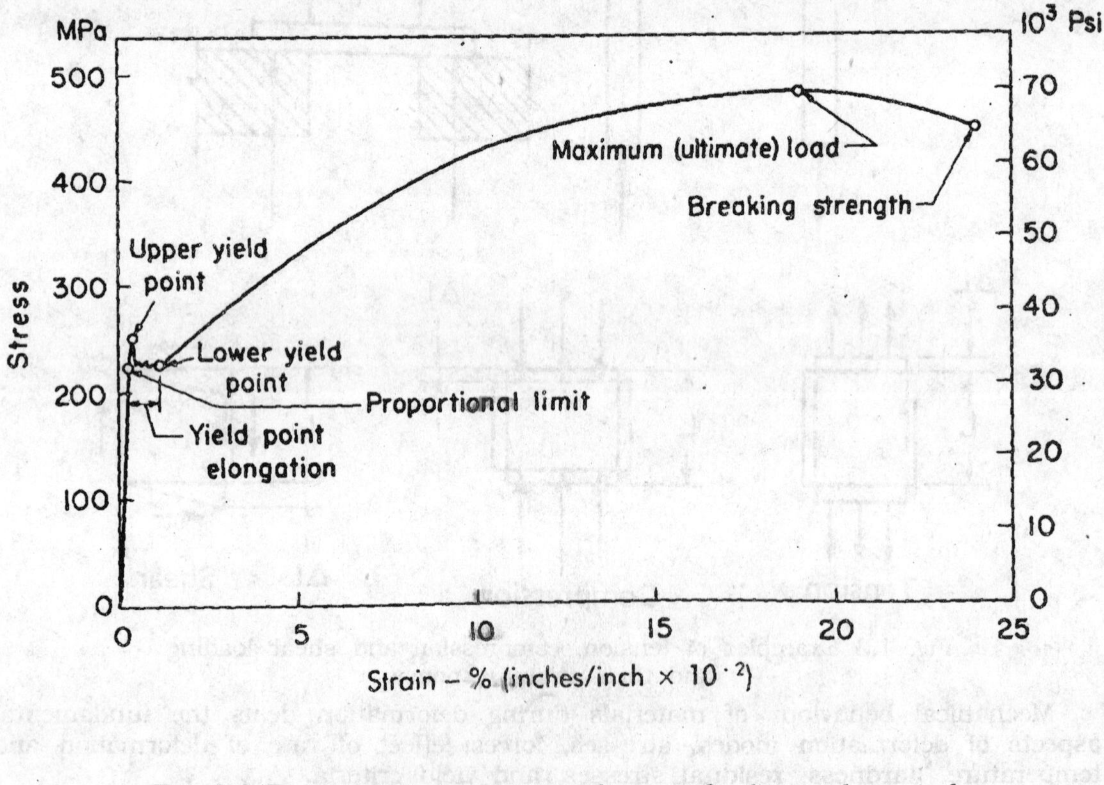

Fig. 1.4 Engineering stress-strain diagram for low carbon steel

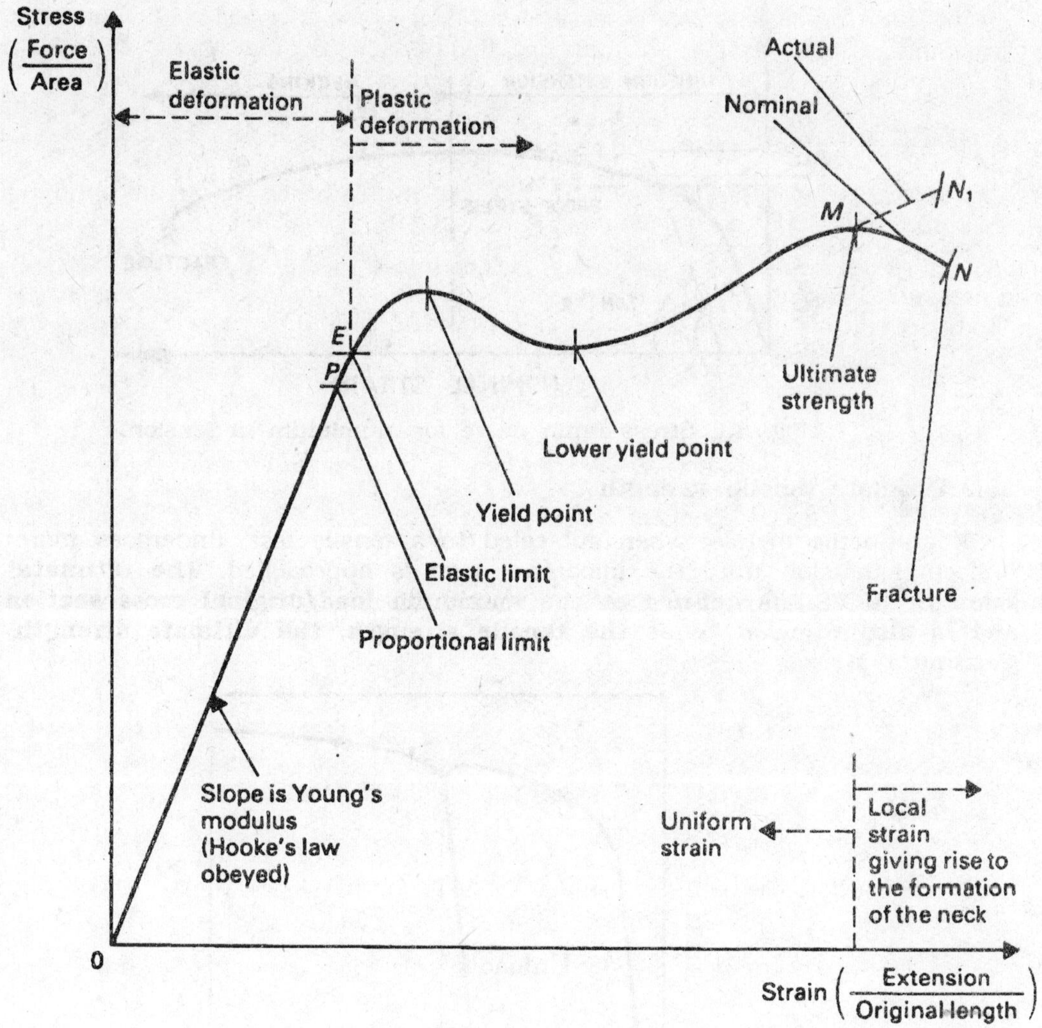

Fig. 1.5 Stress-strain curve

1.5.1 Yield point

In the case of mild steel (and certain other metals) there is a sharply defined upper yield point after which the stress drops abruptly and the specimen extends up to about 1 to 2% strain under an approximately constant lower yield stress.

In contrast, pure aluminium does not exhibit this kind of behaviour at all. Fig. 1.6 shows a typical nominal stress-strain curve for aluminium, from which it can be seen that there is no well defined yield point. Many other non-ferrous metals such as copper, brass and titanium as well as many alloy steels (notably stainless steel) also possess this type of nominal stress-strain curve.

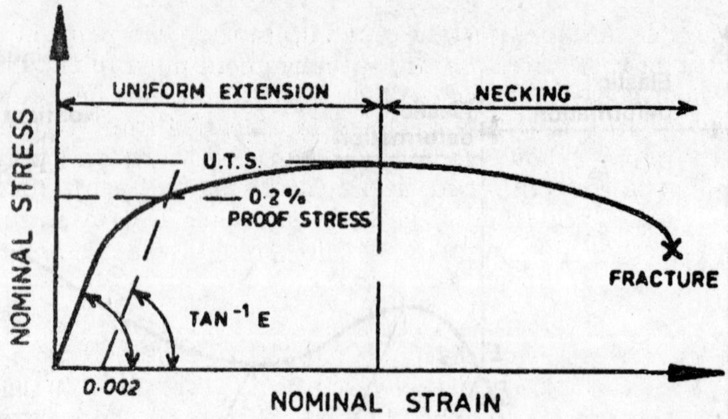

Fig. 1.6 Stress-strain curve for aluminium in tension.

1.5.2 Ultimate tensile strength

Most ductile metals, when subjected to a tensile test, undergoes more or less uniform extension until the maximum load is approached. **The ultimate tensile strength (U.T.S.) is defined as the maximum load/original cross sectional area and is also referred to as the tensile strength, the ultimate strength or the maximum stress.**

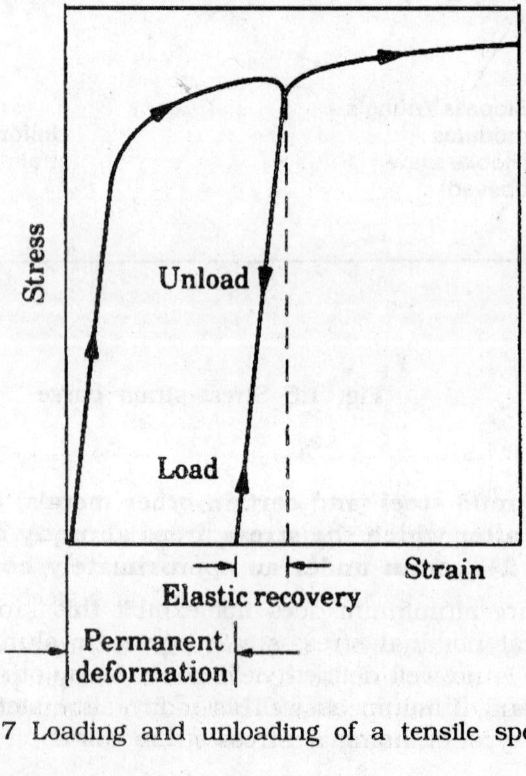

Fig. 1.7 Loading and unloading of a tensile specimen.

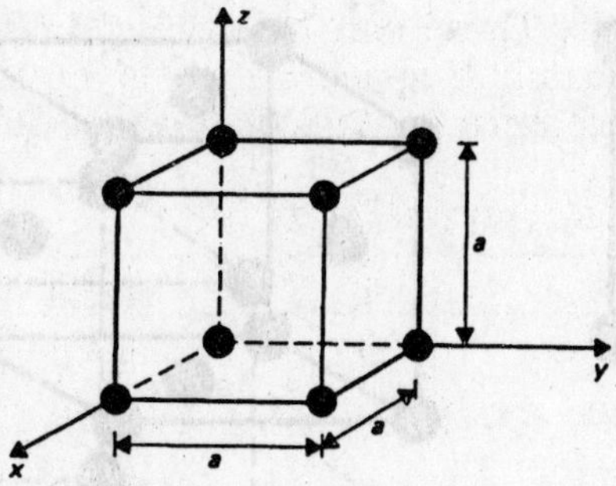

(a) Simple cubic cell. The corners of unit cells are at 'like' positions in the crystal: $a = a = a$. Axes are at right angles.

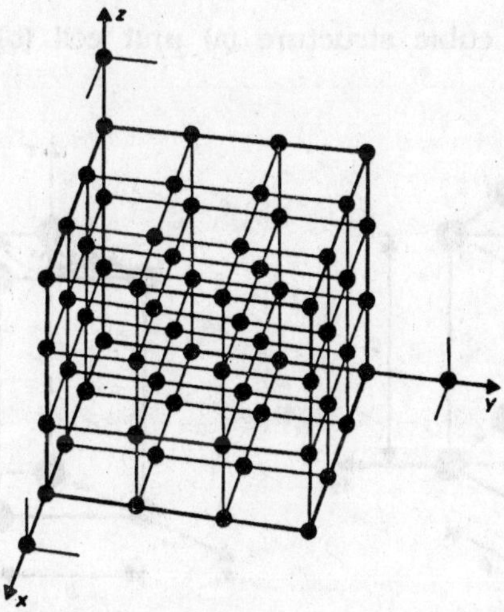

(b) Cubic lattice. Space is divided by three sets of equally spaced parallel planes. The x, y, z reference axes are mutually perpendicular. Each point of intersection is equivalent.

Fig. 2.3 Simple cubic structure.

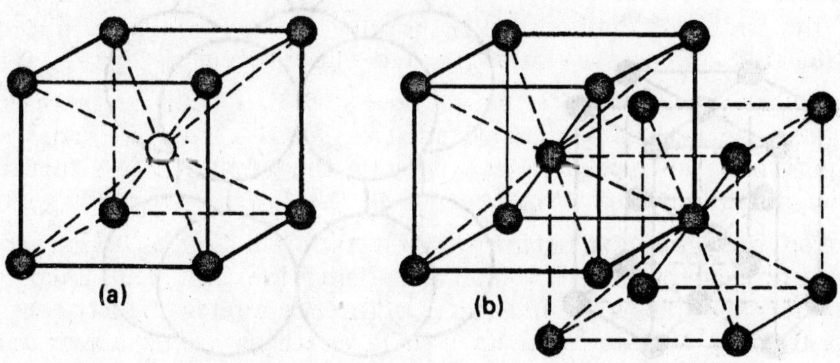

Fig. 2.4 Bode centered cubic structure (a) unitcell (b) crystal lattice.

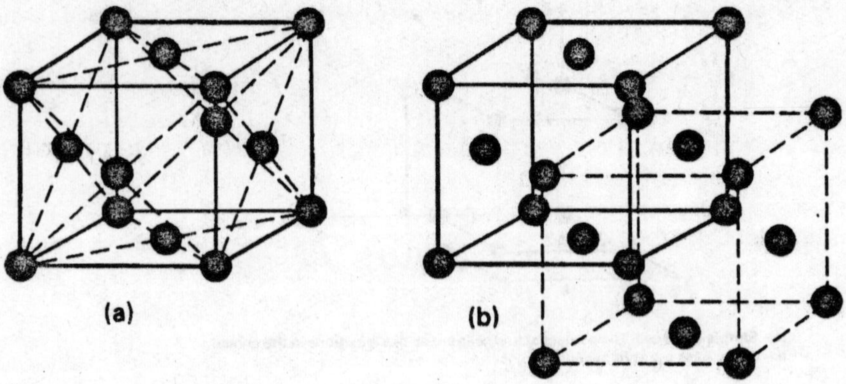

Fig. 2.5 Face centered cubic structure (a) unit cell (b) crystal lattice

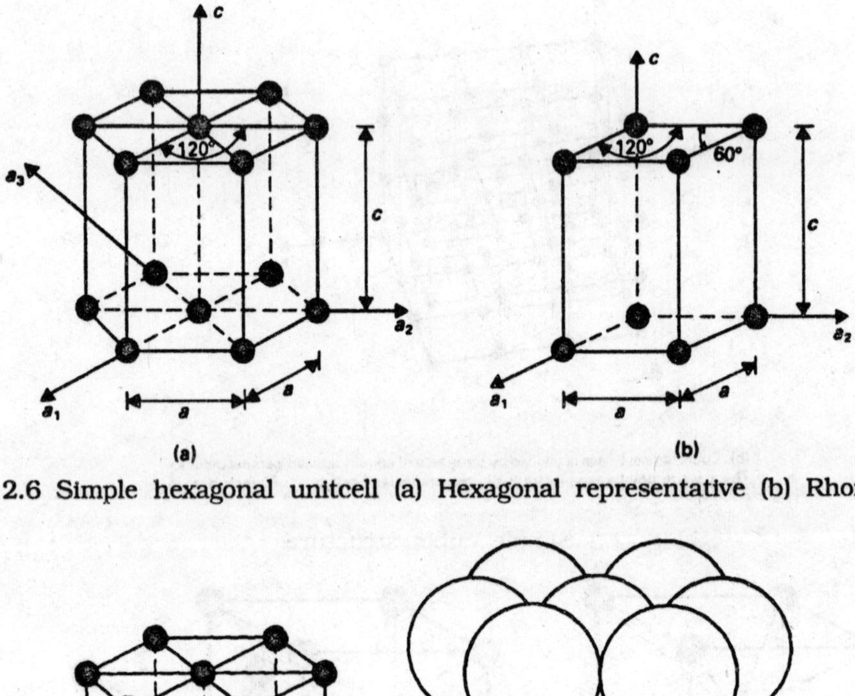

Fig. 2.6 Simple hexagonal unitcell (a) Hexagonal representative (b) Rhombic

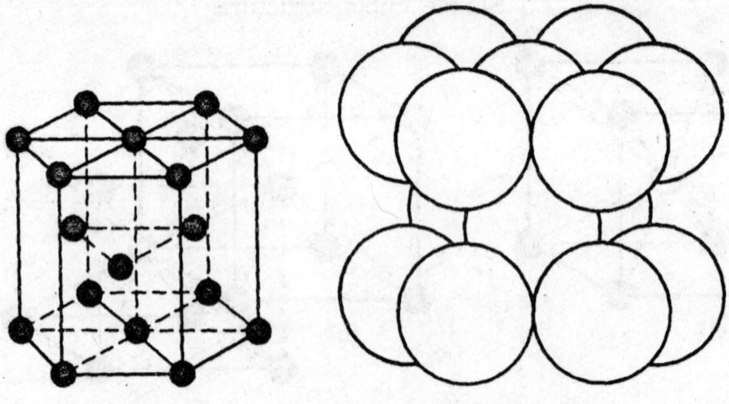

Fig. 2.7 Close packed hexagonal structure

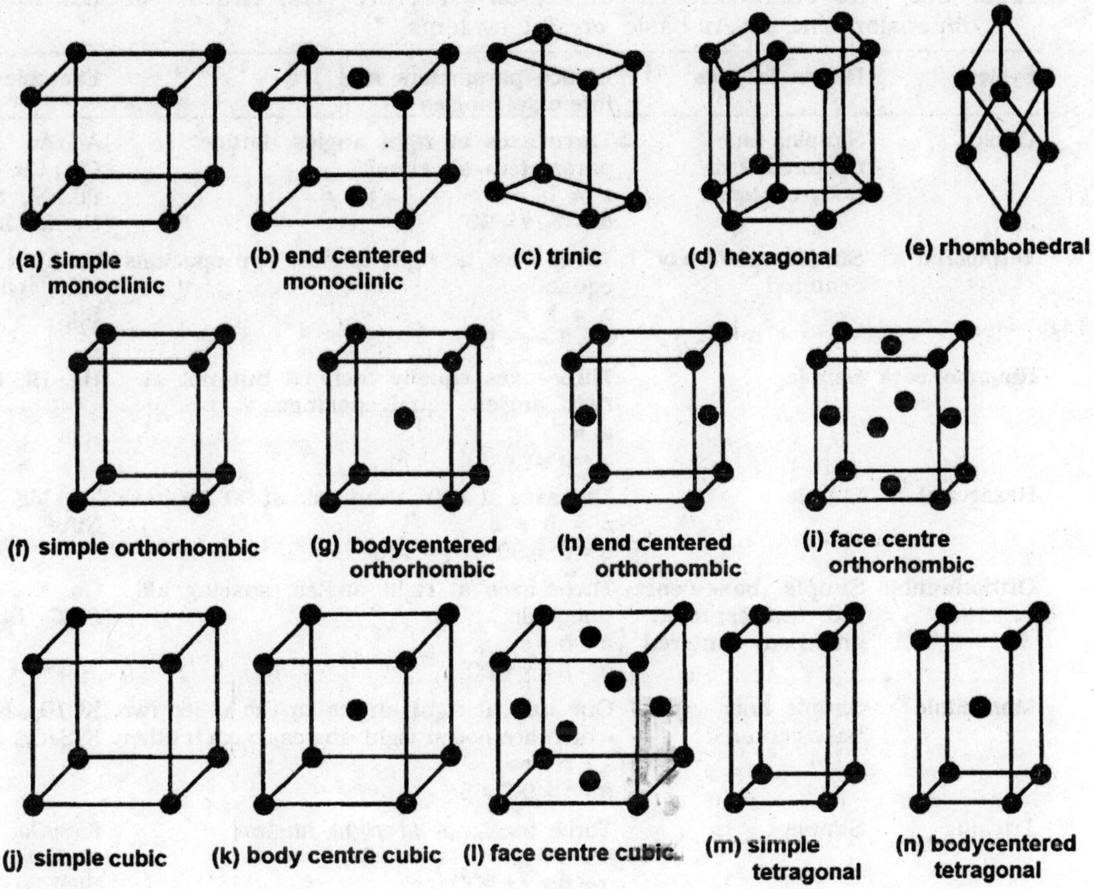

Fig. 2.8 Space lattices. These 14 Bravais lattice continue in three dimensions, each indicated point having identical surroundings.

categories shown in Fig. 2.8. These crystal structures can also be defined by reference to their geometry, as shown in Table 2.2. These are the Bravais space lattices. Of these not more than six are met with in metallic structures and in fact only three really common.

These two are equivalent with a ≠ c, a basal angle of 120° and a vertical angle of 90°.

Table 2.2. The characteristics of crystal structure [The Bravais lattices in these dimensions in seven basic crystal systems].

System	Bravis lattices	Lattice parameters and interaxial angles	Examples
Cubic	Simple, face centered and body-centered	Three axes at right angles, lattice parameters all equal: $a = b = c$ $\alpha = \beta - \gamma = 90^\circ$	Ar, Au, Ag, Cu, Ca, Al, Pb, Ni, Si, NaCl, Lif
Tetragonal	Simple and body centered	Three axes at right angles, two spacings equal: $a = b \neq c$ $\alpha = \beta = \gamma = 90^\circ$	In, TiO_2, KIO_4, white Sn
Rhombohedra	Simple	Three axes equally inclined but not at right angles, equal spacings: $a = b = c$ $\alpha = \beta = \gamma \neq 90^\circ$	Hg, Bi, Sb
Hexagonal	Simple	Two axes at 120°, third axis at 90° to these: $a = b \neq c$ $\alpha = \beta \neq 90^\circ$; $\gamma = 120^\circ$	Zn, Mg, Cd, NiAs
Orthorhombic	Simple, base-centered, face-centered and body-centered	Three axes at right angles, spacing all unequal: $a \neq b \neq c$ $\alpha = \beta = \gamma = 90^\circ$	Ga, I, Fe_2S, Fe_3C, $BaSO_4$
Monoclinic	Simple and base-centered	One axis at right angles to the other two, which are not at right angles to each other: $a \neq b \neq c$ $\alpha = \gamma = 90^\circ \neq \beta$	$KC1O_3$, KNO3, $K_2S_4O_6$, As_2S_4
Triclinic	Simple	Three asex not at right angles: $a \neq b \neq c$ $\alpha \neq \beta \neq \gamma \neq 90^\circ$	K_2S_2Oa, Al_2SiO_5, $B(OH)_3$

2.4 ATOMIC RADIUS

Atomic radius is defined as half the distance between the nearest neighbours in the crystal structure of a pure element. Atomic radius is denoted by r and expressed in terms of the cube edge elements *a*.

(1) Simple cubic structure

In this structure, atoms touch other along the lattice as shown in Fig. 2.9

$$a = 2r$$

$$r = \frac{a}{2}$$

$$\text{and } a^2 = 4\ r^2$$

(2) BCC structure

The atoms touch each other along the diagonal of the cube as shown in Fig. 2.10.

Diagonal of the structure = 4r

$(\text{Body diagonal})^2$ = Sum of the square of all the three sides

$$(4r)^2 = a^2 + a^2 + a^2$$

$$\therefore\ r = \frac{a\sqrt{3}}{4}$$

$$\text{and } a^2 = \frac{16\,r^2}{3}$$

(3) FCC structure

Atoms within this structure touch along the diagonal of any face of the cube as shown in Fig. 2.11.

Diagonal of the structure = 4 r

$(\text{Body diagonal})^2$ = Sum of the squares of the two sides

$$(4r)^2 = a^2 + a^2$$

$$r = \frac{a}{2\sqrt{2}}$$

$$\text{and } a = 8\,r^2$$

2.5 DENSITY OF CRYSTAL

This is defined as

$$\text{Density of the crystal} = \frac{\text{Mass of unit cell}}{\text{Volume of unit cell}}$$

$$\text{Mass of unit cell} = \frac{\text{Atomic mass}}{\text{Avogadro number}} \times \text{Number of atoms per unit cell}$$

$$= \frac{M}{N_A} \times n$$

2.6 CHARACTERISTICS OF CRYSTAL STRUCTURES

For metals there is an interrelationship between the melting point, the elastic modulus (which is a measure of the bending) and the strength (resistance to deformation).

This is shown by the variation in values of a particular property for different metals, and values of the parameters which describe the crystal structures.

2.6.1 Coordination number

For a particular atom in a regular repeating crystal structure, the coordination number is the number of nearest (or first) neighbours of the atom.

2.6.2 Atomic Packing factor

This is the fraction of the unit cell which is occupied by the atoms comprising that structure. The metal atoms are considered to be 'hard balls' of radius r for a cubic structure with unit cell dimensions *a*.

(1) Simple cubic cell

Fig. 2.9 shows simple cubic cell.

$$\text{Atomic packing factor} = \frac{\text{Volume of atoms}}{\text{Volume of unit cell}}$$

$$= \frac{(4/3)\,\pi\, r^3}{a^3}$$

$$= \frac{(4/3)\,\pi\, r^3}{(2r)^3} = 0.52.$$

(2) Body centered cubic

The lattice constant (a) is related to the atomic radius (r) as shown in Fig. 2.10 such that.

This structure has two atoms per unit cell, one atom at the centre and each corner atom is shared by eight unit cell.

$$(\text{Body diagonal})^2 = \text{Sum of the square of all the three sides}$$

$$(4r)^2 = a^2 + a^2 + a^2$$

$$\therefore a = \frac{4r}{\sqrt{3}}$$

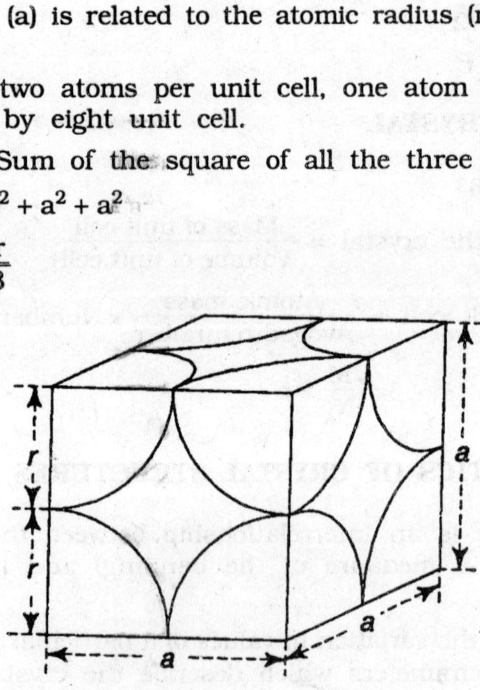

Fig. 2.9 Simple cubic cell.

$$\text{Atomic packing factor} = \text{Number of atoms} \times \frac{\text{Volume of one atom}}{\text{Volume of unit cell}}$$

$$= \frac{2\left(\frac{4}{3}\right)\pi r^3}{(4r/\sqrt{3})^3} = 0.68$$

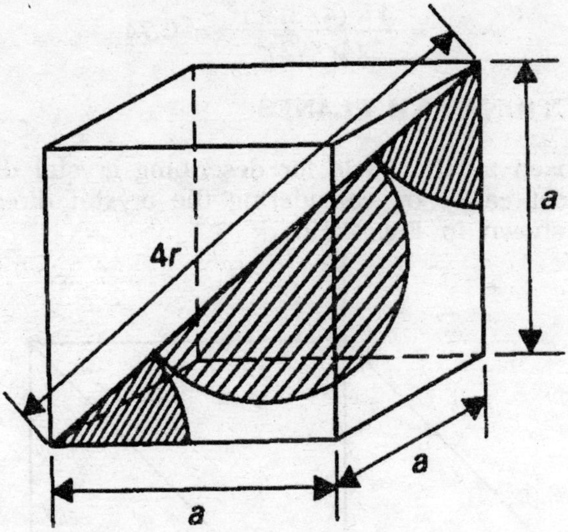

Fig. 2.10 BCC unit cell

(3) Face-centered cubic

The structure has four atoms per unit cell; each corner atom is shared by eight unit cells and each face centered atom is shared by two unit cells.

Fig. 2.11 shows face centered cubic unit cell. A section across the face of a unit cell.

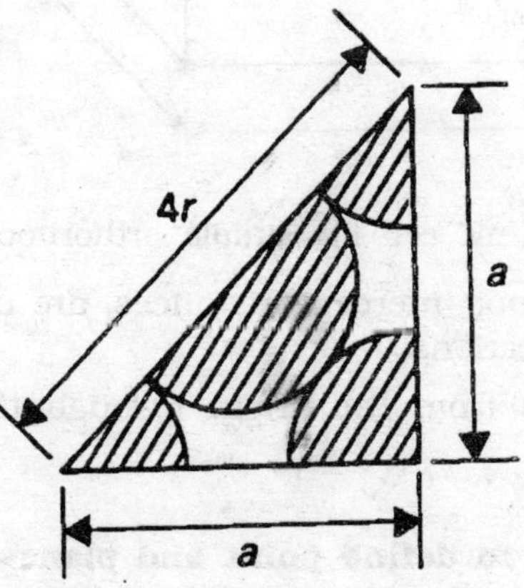

Fig. 2.11 FCC unit cell

$$(4r)^2 = a^2 + a^2$$

$$\therefore a = \frac{4r}{\sqrt{2}}$$

$$\text{Atomic packing factor} = \text{Number of atoms} \times \frac{\text{Volume of atoms}}{\text{Volume of unit cell}}$$

$$= \frac{4 \times (4/3)\,\pi\, r^3}{(4r/\sqrt{2})^2} = 0.74$$

2.7 CRYSTAL DIRECTIONS AND PLANES

The unit cell is used as the basis for describing crystal directions and planes. This can be seen most easily by considering the crystal directions on the simple orthorhombic lattice shown in Fig. 2.12.

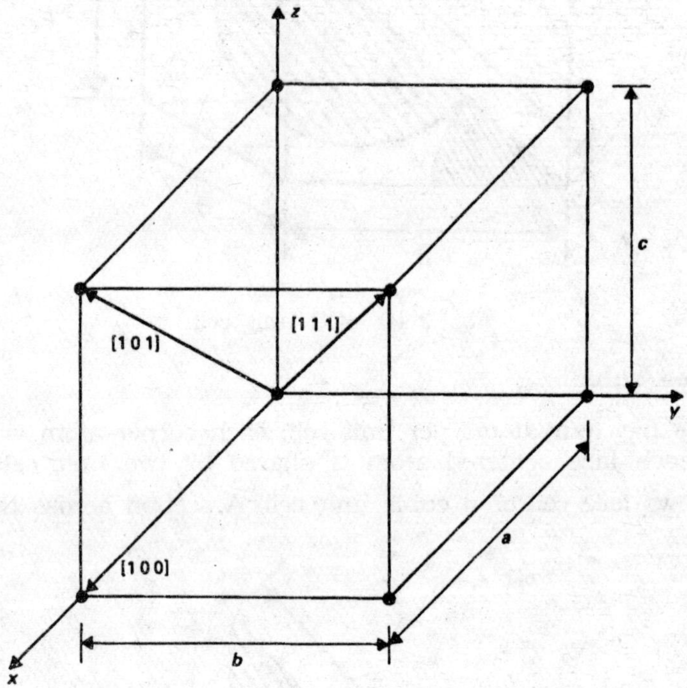

Fig. 2.12 Unit cell for simple orthorhombic lattice.

Unit cell distances, and not intercept distances, are used ($a \neq b \neq c$). The crystal axes are used as base directions.

The direction [1 1 2] is from the origin through the top face.

2.7.1 Miller Indices

Miller indices are used to define point and planes within a crystal structure. The atomic spacing is the lattice parameter with respect to an arbitrary origin. **A**

point is defined in terms of its coordinates, and a plane in terms of the reciprocals of its intercepts with the axes.

The (0 1 0) planes are parallel to the x and z axes, and the (1 1 1) planes cut the three crystal axes as shown in Fig. 2.13.

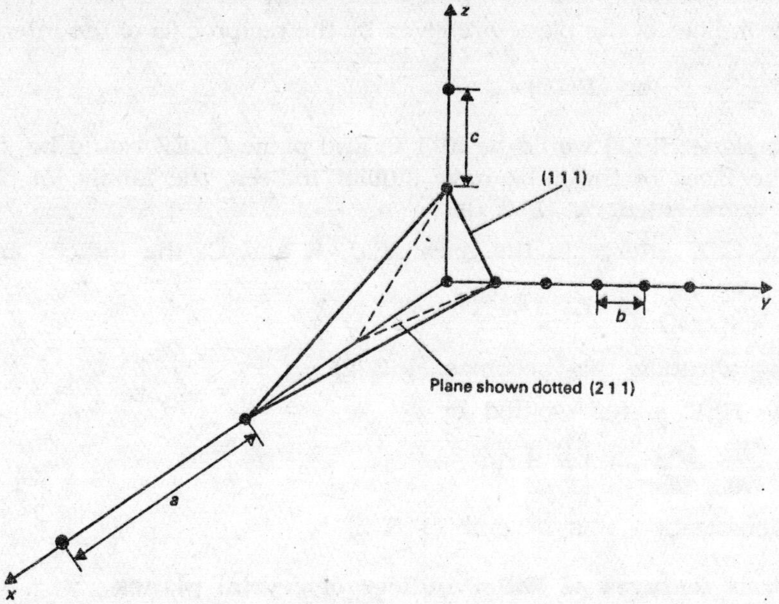

Fig. 2.13. Definition of crystal planes using Miller indices

With reference to a unit cell assuming a plane cuts the axes at 3/2, 2, 2. Take reciprocals of these intercepts, obtaining 2/3, 1/2, 1/2, then rationalize to obtain (4 3 3). These are the Miller indices of the plane.

With reference to Fig. 2.14, where OA, OB and OC represent the sides of a unit cell, then

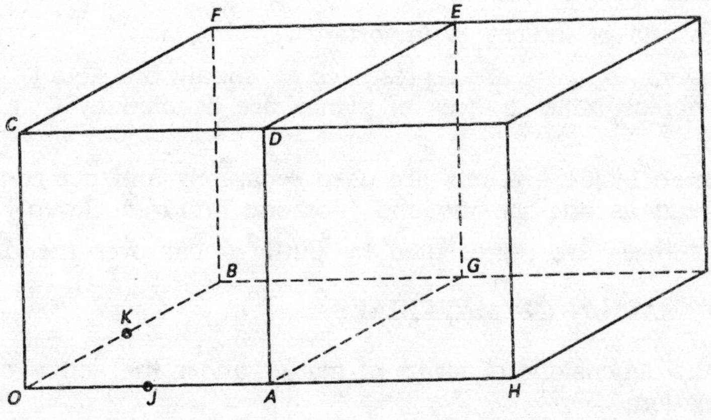

Fig. 2.14 Unit cell used to define the Miller indices of a plane.

OH = 2 x OA

OJ = 1/2 x OA

OK = 2/3 x OB

The plane ADEG intercepts the axes at A, ∞.

The Miller indices of the plane are given by the reciprocals of the intercepts, namely

$$\frac{OA}{OA} \frac{OB}{\infty} \frac{OC}{\infty} \text{ or } (1\,0\,0)$$

Similarly, plane BFEG would be (0 1 0) and plane CDEF would be (0 0 1). Planes parallel to the faces of the cube have similar indices; the family, of planes parallel to ADEG is represented by (1 0 0).

The plane CJK intercepts the axes at J, K and C, the indices are given by

$$\frac{OA}{OJ} \frac{OB}{OK} \frac{OC}{OD} \text{ or } (2\ 3/2\ 1)$$

After rationalization this becomes (4 3 2)

The plane HKC is represented by

$$\frac{OA}{OH} \frac{OB}{OK} \frac{OC}{OC} \text{ or } \left(\frac{1}{2}\ \frac{3}{2}\ 1\right)$$

After rationalization this become (1 3 2)

2.7.2 Important features of Miller indices of crystal planes

A few important features of Miller indices of crystal planes are listed below.

(1) Miller indices do not only define a particular plane but a set of parallel planes.

(2) A plane which is parallel to any one of the principal axes has an intercept of infinite (∞) and the Miller index is therefore zero.

(3) All equally spaced parallel planes with a particular orientation have the same index number. [h k l]

(4) Only the ratio of indices is important.

(5) The directions in space are represented by square brackets [ ] and the letter x y z, whereas Miller indices of planes are denoted by ( ) and the letters h k l.

(6) The common inside brackets are used separately and not combined. Thus (1 1 1) is read as one-one-one and "not one hundred eleven".

(7) Negative indices are represented by puttig a bar over the digit $(0\ \bar{1}\ 0)$.

2.8 THE DEFORMATION OF MATERIALS

The change in dimensions of forms of matter under the action of applied forces is called deformation.

The metal deformations are of the following two types :

1. Elastic deformation, and
2. Plastic deformation.

2.8.1 Elastic deformation

To a great extent, the mechanical behaviour of materials depends on their crystal structure. In general, observation reveals that the behaviour of metal crystal depends on

(1) the lattice type;

(2) the interatomic forces;

(3) the spacing between planes of atoms; and

(4) the density of atoms on various planes.

If the applied loads are relatively low, the crystal responds by simple stretching or compressing the distance between atoms, as in Fig. 2.15. The lattice unit does not change, and the atoms retain their basic positions. The applied load serves only to disrupt the force balance of the atomic bonds so as to transmit the applied load through the body. When the load is removed balance is restored and the lattice resumes its original size and shape. The response to such loads is elastic in nature. The amount of stretch or compression (strain) is proportional to the applied load or stress.

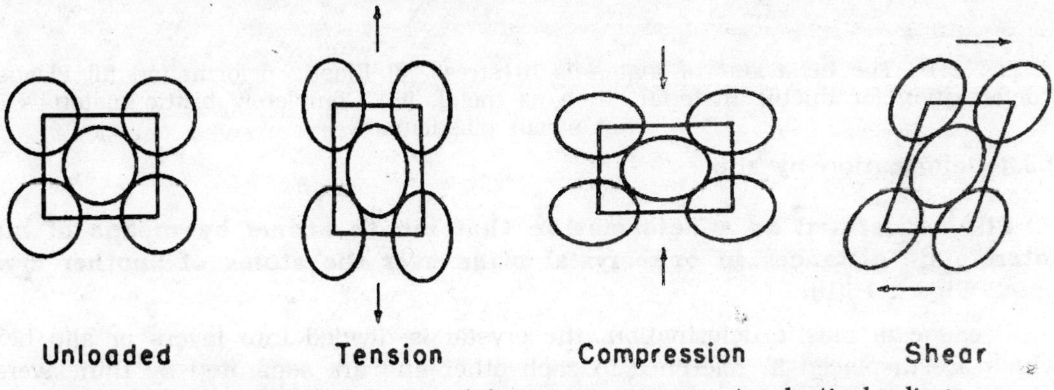

Fig. 2.15 Distortion of a crystal lattice in response to elastic loadings.

2.8.2 Plastic deformation

As the magnitude of applied load is increased, the distortion increases to a point where the atoms must either (1) break bonds to produce a fracture or (2) slide over one another to produce a permanent shift of atom positions. The result is a plastic deformation, where a permanent change in shape occurs without a concurrent deterioration in properties. (Fig. 2.16)

There are two basic mechanism by which plastic deformation may take place in crystal structures : slipping of one plane of atoms over an adjacent plane. The second mechanism of plastic deformation is twinning, in which a portion of the crystal forms structurally a mirror image of itself across the plane of twinning.

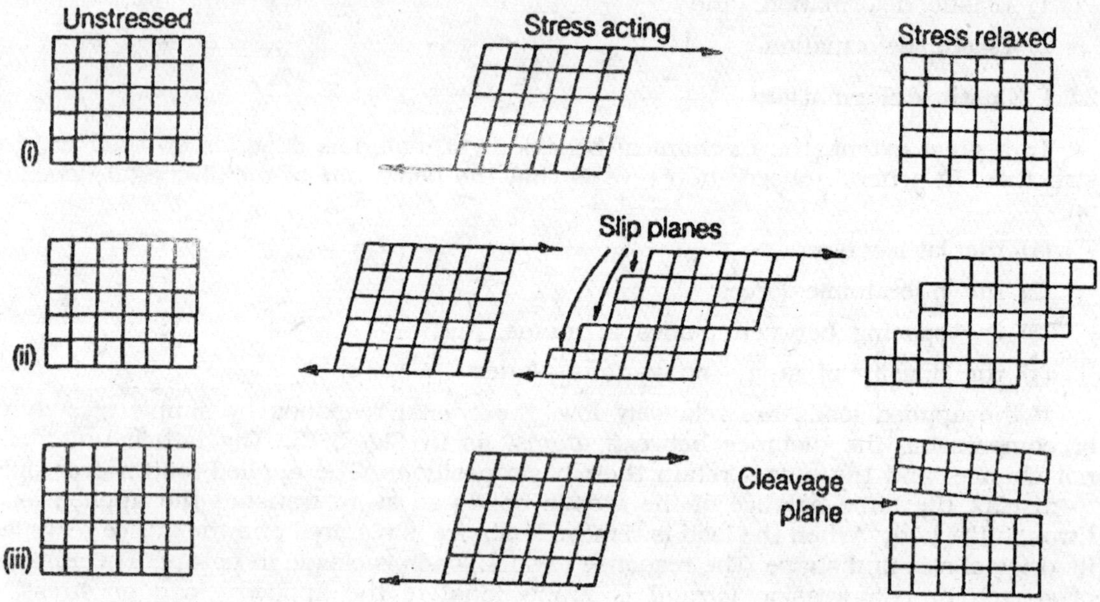

Fig. 2.16 The behaviour of materials in stress. (i) Elastic deformation (ii) Plastic deformation for ductile material, such as metal. (iii) Completely brittle materials do not strain plasticity.

2.8.3 Deformation by slip

Slip is defined as a deformation that moves atoms by means of many interatomic distances in one crystal plane over the atoms of another crystal plane Fig. 2.16(ii).

Because of plastic deformation, the crystal is divided into layers or slip blocks which are displaced in reference to each other and are separated by thin layers in which considerable displacement of atoms has taken place.

These intermediate layers with strongly distorted lattice are called 'slip plane's.

The cylindrical crystal shown in Fig. 2.17 illustrates two aspects of crystalline slip.

(1) the original circular cross section of the crystal becomes elliptical; and

(2) surface makings appear as the slip planes become exposed on the surface of the crystal and thus the outer surface becomes roughened.

Three general observations can be made concerning slip movements in crystals.

(1) Slip begins when the shearing stress acting along the slip plane in the direction of slip exceeds a certain value which is known as the critical shearing stress.

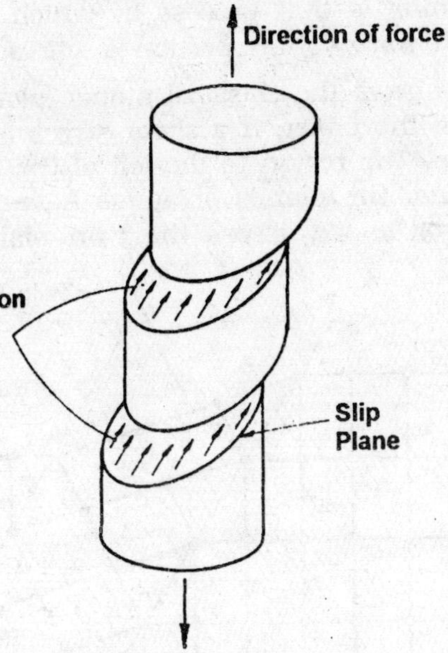

Fig. 2.17 Slip planes and slip directions in a single cylindrical metal crystal.

(2) Values of critical shear stress depend upon the composition of the crystal, on the temperature, and on previous mechanical working, it is lowest for pure metals and decreases steadily with increasing temperature, while previous cold working increasing it quite considerably.

(3) Slip planes are always planes of high atomic densities, while direction of slip along these planes is always the direction of highest atomic density.

2.8.4 Twinning

The second mechanism of plastic deformation is twinning, in which a portion of the crystal forms structurally a mirror image of itself across the plane of twinning. Fig. 2.18 shows twinning in tension.

Fig. 2.18 Twinning in tension

In other words twinning is that process by which a portion of the crystal takes up an orientation which makes that portion a mirror image of the parent crystal.

Fig. 2.19 (a) and (b) shows the classical atomic picture of twinning. The twinning plane is at right angle to the paper. If a shear stress is applied, the crystal will turn about the twinning plane. The region to the left of the twinning plane is undeformed. To the right of this plane, the planes of atoms have sheared in such a way as to make the lattice a mirror image across the twin plane.

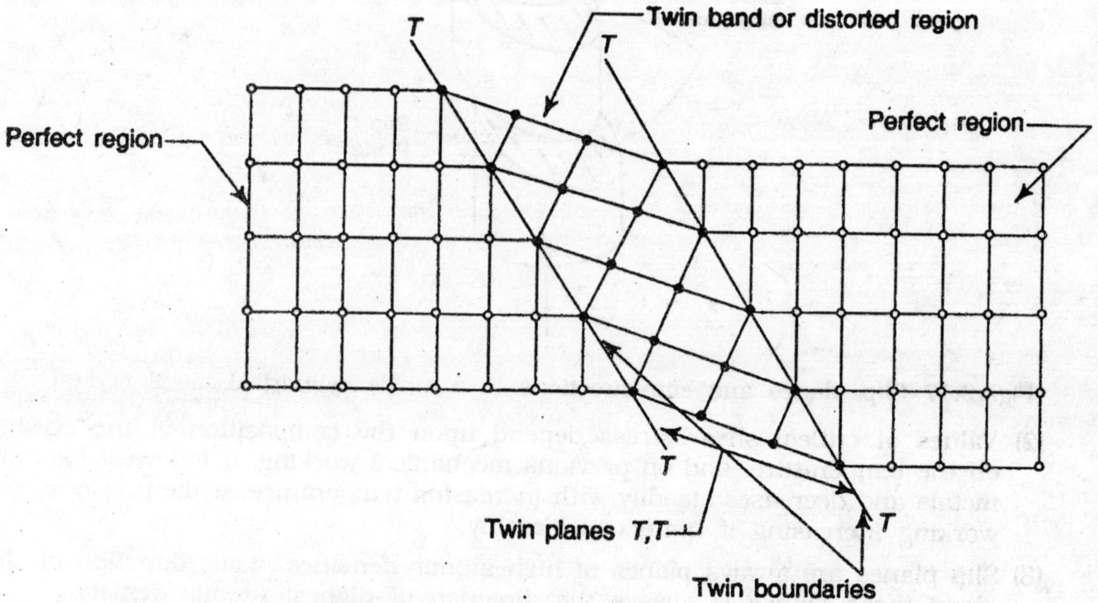

Fig. 2.19a. Classical atomic picture of twinning.

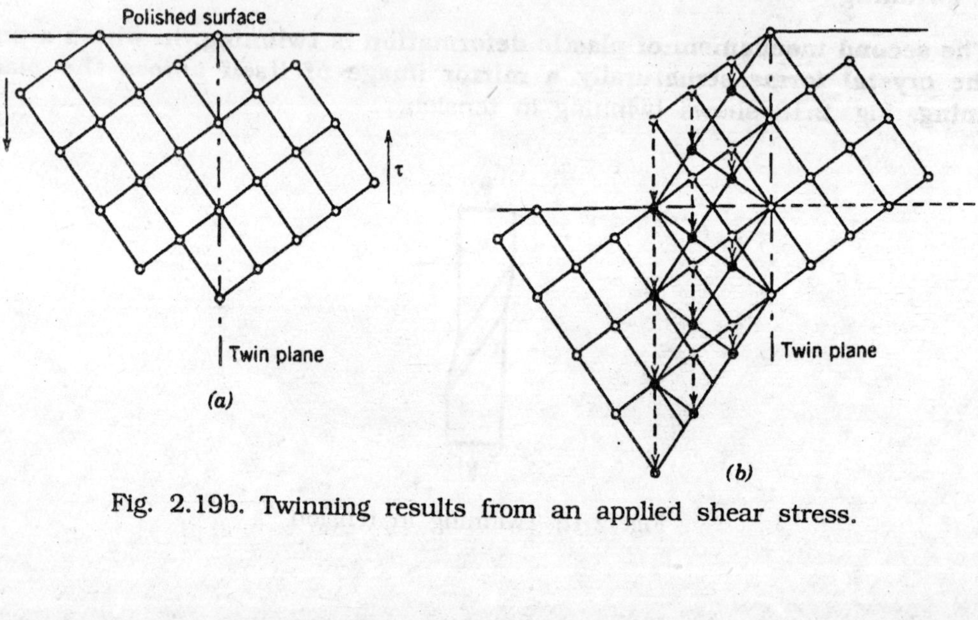

Fig. 2.19b. Twinning results from an applied shear stress.

Twinning does not occur as frequently as slip, it nevertheless is important. Since quite a few metals tend to deform under certain circumstances by twinning.

A twin crystal is formed when one part of a crystal reorients itself so that it becomes crystallographic plane about which twinning occurs is known as the twin plane and is plane of symmetry within the twinning crystal.

Twinning differs from slip in that every plane of atoms suffer some movement and crystallographic orientations of many unit cells are altered.

Twinned crystals are of two kinds.

(1) Mechanical twins or strain twins

Strain twins are produced as an immediate result of applied stress in the manner as outlined.

Strain twins are produced by mechanical deformation. It is in BCC or HCP metals under conditions of rapid rate of loading (shock loading) and decreased temperature.

(2) Annealing twins

Annealing twins only appear in annealed metals that have been subjected to prior mechanical working at ordinary temperature. These are more common than strain twins.

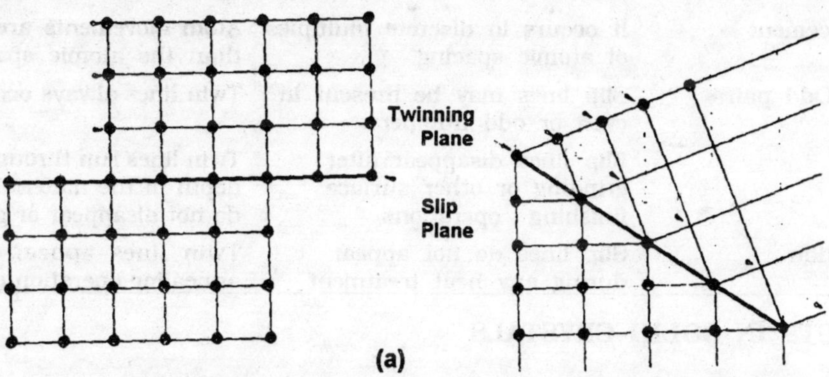

(a)

Fig. 2.20 (a) Relative atom movements in slip and twinning

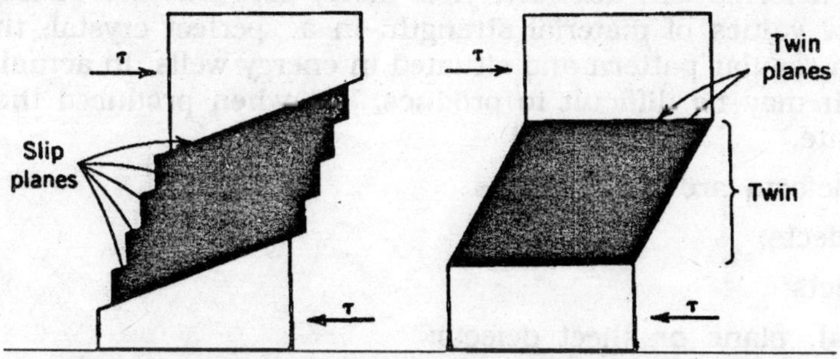

Fig. 2.20 (b) Deformation by slip and twinning.

It is most important to note that neither slip planes nor twin planes extend beyond the grain boundaries of the individual crystals in which they occurs (Fig. 2.20 (a) and (b). Thus in polycrystalline aggregates both slip and twinning are intercrystalline, indicating that the grain boundaries are stronger than grain themselves.

2.8.5 Difference between slip and twinning

The difference between slip and twinning are given in table 2.3.

Table 2.3. Difference between slip and twinning

SN	Title	Slip	Twinning
1	Crystal Orientation	The orientation of the crystal above and below the slip	Orientation difference occurs across the plane
2	Minimum stress limit	Slip begins when shearing stress on the slip plane in the slip direction reaches a threshold value called the critical shear stress.	There is no limit for critical resolved shear stress for twinning
3	Deformation time	Many million second to form a band.	It takes place in a few mincroseconds.
4	Affected planes	It occurs over relatively wide planes	Every atomic plane is involved in deformation.
5	Displacement	It occurs in discrete multiples of atomic spacing	Atom movements are much less than the atomic spacing.
6	Even/Odd pairs	Slip lines may be present in even or odd numbers.	Twin lines always occur in pairs.
7	Lines	Slip lines disappear after grinding or other surface finishing operations	Twin lines run through the whole depth of the material and hence do not disappear or grinding etc.
8	Operation	Slip lines do not appear during any heat treatment	Twin lines appear during the annealing operation of materials.

2.9 DEFECTS IN SOLID CRYSTALS

Defects in crystals are important because they determine the mechanical behaviour, such as plastic deformation, ductility, yield stress and fracture stress. They are also the cause of low values of material strength. In a perfect crystal, the atoms would be arranged in a regular pattern and situated in energy wells. In actual crystal, defects can occur which may be difficult to produce, but when produced they are relatively easy to propogate.

Structural defects are classified as

(1) point defects;

(2) line defects;

(3) interfacial, plane or sheet defects;

(4) bulk defects.

2.9.1 Point defects

In its simplest form a point defect is a vacancy caused by a missing atom, and it may involve more than one site vacancy. Defects may occur due to imperfect packing of the atoms during crystallization, or at high temperatures thermal vibration of the atoms may cause movement from positions of low energy. In compounds this can be a cation or anion vacancy.

A Schottky imperfection is a defect composed of both an anion and a cation vacancy, occurring in a compound which must maintain a charge balance [Fig. 2.21].

Within a crystal structure there may be foreign atoms which do not belong to the given lattice; these are impurities. They can occur either in lattice sites as **substitutional impurities** or between lattice sites as **interstitial impurities**. Foreign atoms which are small compared to the matrix atoms are often found as interstitial impurities. Larger atoms are most common as substitutes, otherwise, they can cause atomic distortion.

A Frankel defect is the displacement of an ion from its lattice position into an interstitial site [Fig. 2.21].

Addition energy is required to transfer atoms to new positions, and close-packed structures have more vacancies and Schottky defects than interstitial and Frankel defects.

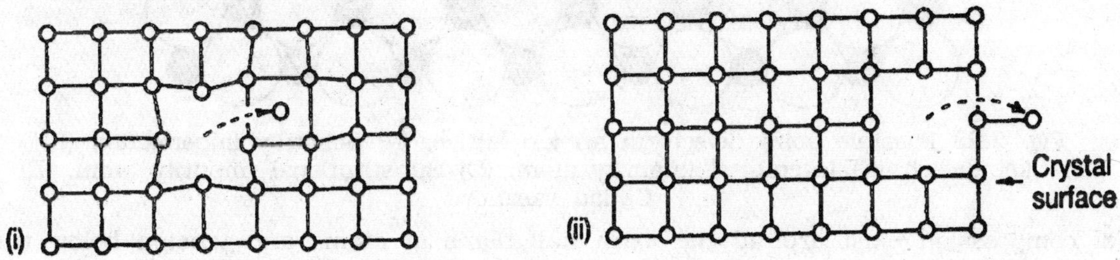

Fig. 2.21 (i) A Frankel defect is a vacancy produced when an atom moves. (ii) A Schottky defect is a vacancy formed in a similar manner but the displaced atom takes up a lattice position on the surface of the crystal.

The main types of point defects are shown in Fig. 2.22.

2.9.2 Line defects

These are generally edge dislocations or screw dislocations. It is possible for mixed dislocations to occur (e.g. the dislocation loop) but these are more difficult to describe and are usually represented as superpositions of edge and screw dislocations. A graphical representation has been devised by Bergers and modified by Frank (the Frank-Burgers circuits or vectors).

An Edge dislocation is due to an extra plane of atoms within a crystal structure, and this is represented diagrammatically by the symbol ( ) Regions

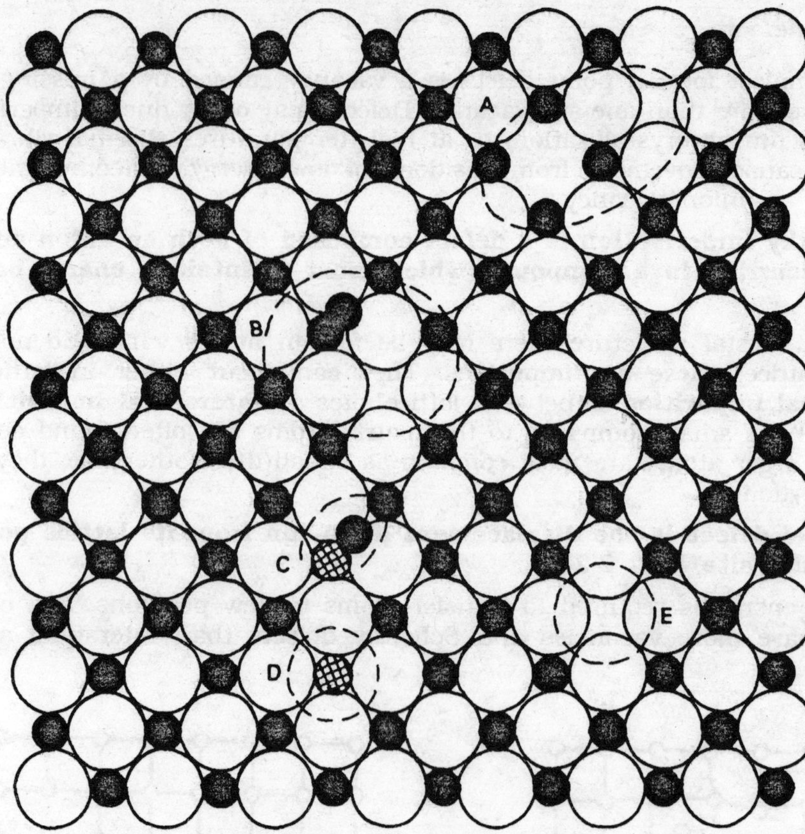

Fig. 2.22 Possible point defects in an ion lattice. (A) Schottky imperfection (B) Frenkel defect. (C) Interstitial impurity atom. (D) Substitutional impurity atom. (E) Cation vacancy.

of compression exist around the extra half plane of atoms and tension below the extra half plane of atoms. These regions are created within the crystal structure, resulting in a net increase in energy along the dislocation. **An edge dislocation has its displacement perpendicular to the dislocation line.** [Fig. 2.23]

A screw dislocation (Symbol (   )) has its displacement parallel to the defect. The associated shear stresses also cause a net increase in energy. [Fig. 2.24]

Edge dislocations occur during crystal formation when adjacent parts of the growing crystal are misaligned, so that an extra row of atoms is introduced or eliminated. With a screw dislocation, extra atoms can be added to the step of the screw.

The geometry of linear dislocations are given in Fig. 2.25.

Fig. 2.26 illustrate the movement of an edge dislocation across the crystal lattice under a shear stress and permanent deformation occurs. The dislocation moves through the array of atoms without whole sale movement of planes of atoms past each other.

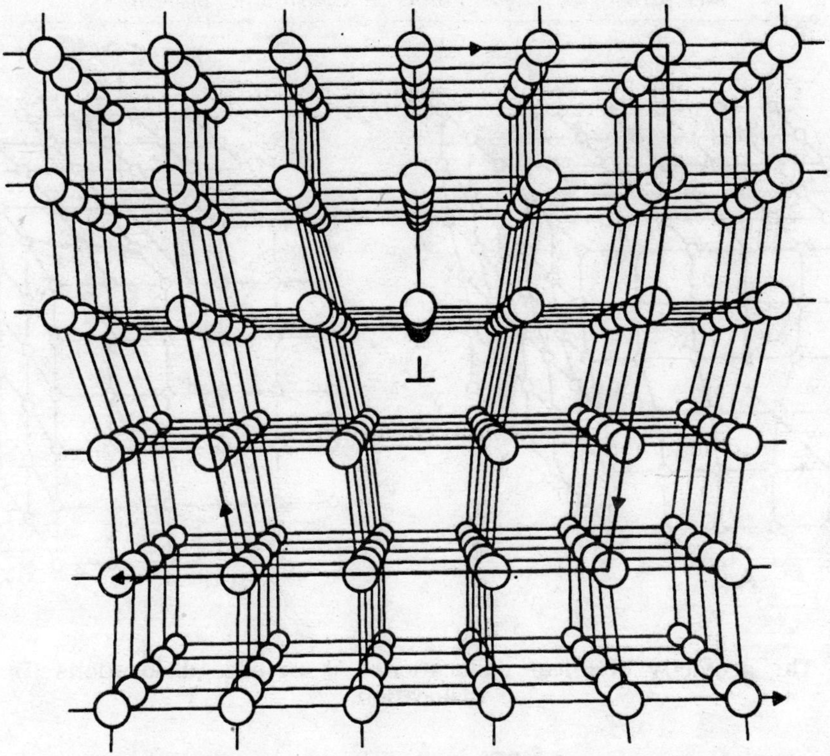

Fig. 2.23 Atom arrangement in an edge dislocation. A linear defect occurs at the edge of an extra plane of atoms.

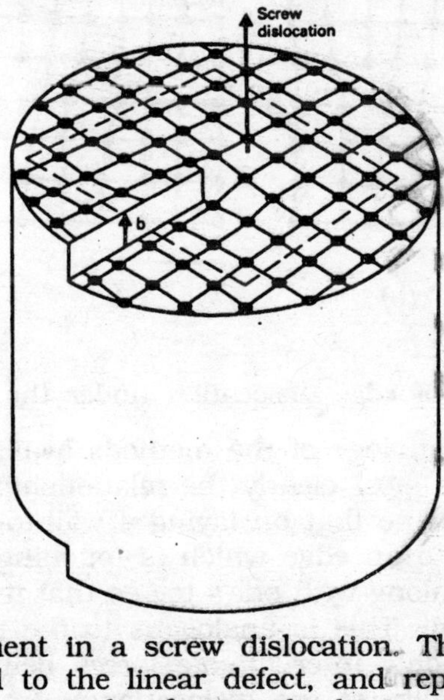

Fig. 2.24 Atom arrangement in a screw dislocation. The displacement vector 'b' (Berger Vector) is parallel to the linear defect, and represents the direction and magnitude of atom displacement.

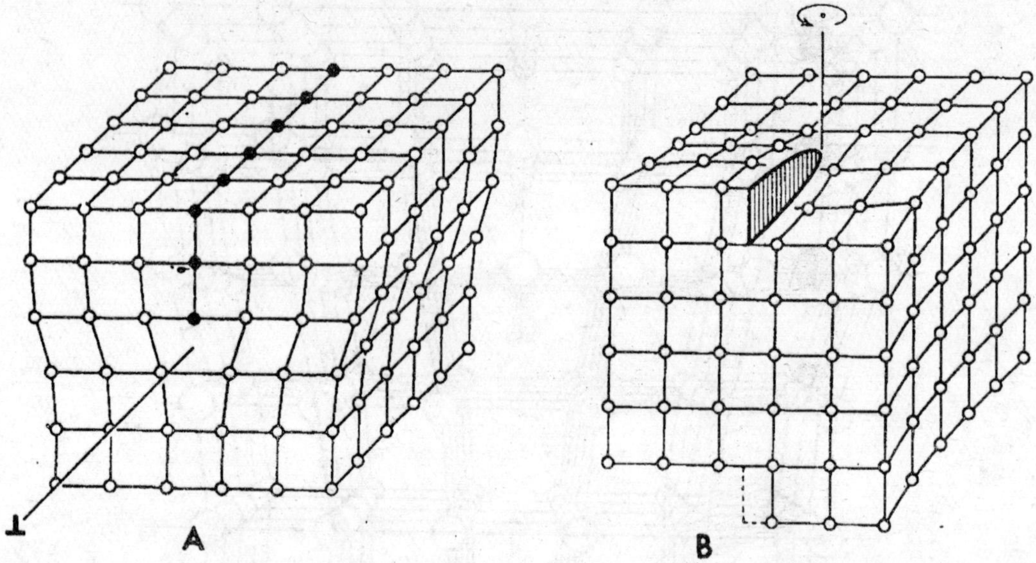

Fig. 2.25 The geometry of linear dislocations (A) an edge dislocations (B) a screw dislocation.

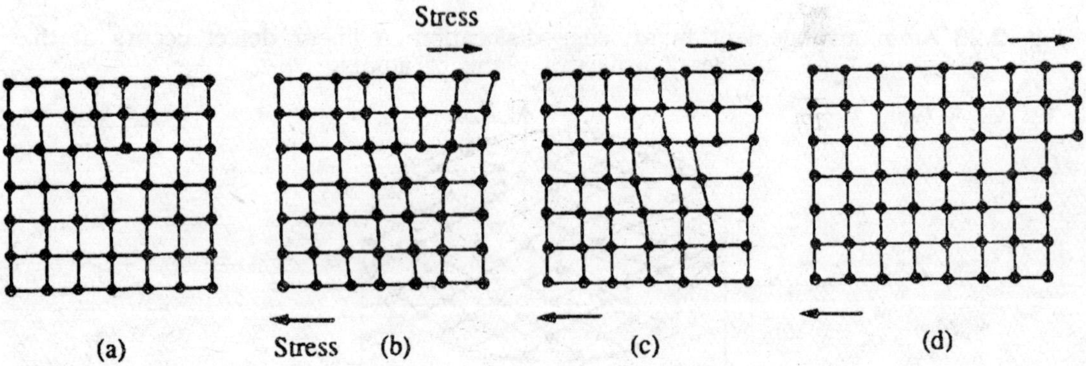

Fig. 2.26 Movement of edge dislocation under the influence of stress.

Prof. N. Mott's classical analogy of the methods available for smoothing wrinkles from a heavy carpet explains most clearly the relationship between slip and the force necessary to produce it. Imagine that on laying a wall-to-wall carpet a small wrinkle remains in the carpet near to an edge which is touching a wall. [Fig. 2.27 (i)]. Very little effort coax the wrinkle along with one's toe so that it moves, step by step, across the floor (Fig. 2.27 (ii) to (iv)]. This is analogous to the progressive movement of an edge dislocation where the only inter atomic forces being overcome at any instant are those acting in the locality of the dislocations.

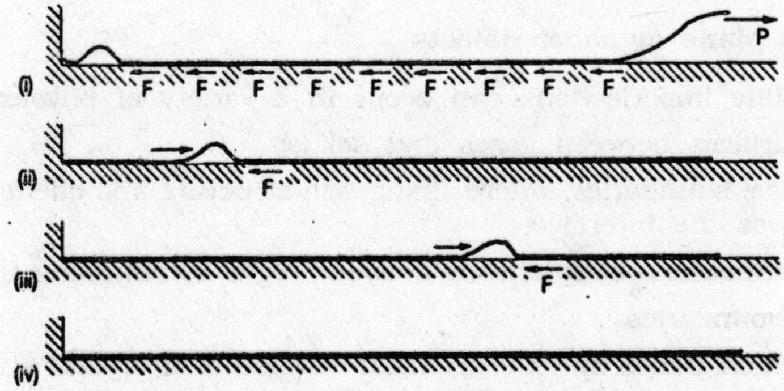

Fig. 2.27 Mott's analogy of the carpet

Similarly analogy for earth worm which moves starts at the trail and moves forward the head.

Fig. 2.28 shows the form of a screw dislocation and its movement through the array of atoms under the action of stress. With an edge dislocation the line of dislocation is at right angles to the slip plane; with a screw dislocation the line of dislocation is parallel to the slip plane.

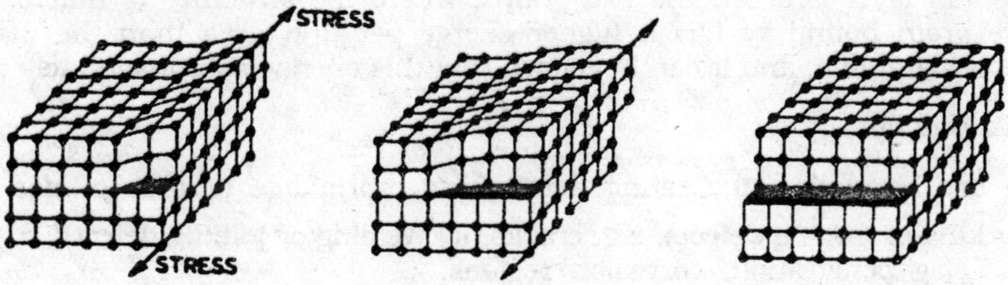

Fig. 2.28 The movement of screw dislocation.

Burger circuits

In practice, dislocations are often neither straight lines, at right angles to, or parallel to, the slip plane but curved lines, which can, however, be considered to be a combination of edge and screw dislocations (Fig. 2.29).

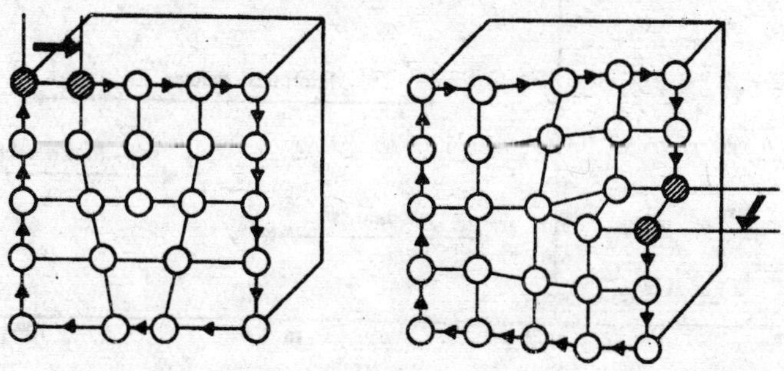

Fig. 2.29 Burgers circuits for edge and screw dislocations

2.9.3 Interfacial, plane or sheet defects

These crystalline imperfections can occur in a variety of physical situations :

(a) at free surfaces between gases and solids;

(b) at interphase boundaries, where changes in structure and chemical composition occur across the interface;

(c) domain boundaries, at interfaces between regions containing the same atoms.

(d) at grain boundaries.

The atoms which make up the external surface of a crystal structure possess more energy than the internal atoms because these atoms are not entirely surrounded. An example of this state is the spherical shape assumed by liquid droplets possessing the minimum surface area, and surface energy, per unit volume.

Grains are individual crystals which combine to make up the solid metallic phase, and which exist in a variety of orientations. The shape of a grain is normally determined by the surrounding grains. The atoms within a grain have the same orientation and packing, which is determined by the lattice arrangement. At the boundary between two adjacent grains the atoms cannot align perfectly with both crystal lattices. There is a transition layer between the two grains, where the structure is that of neither grain. The grain boundary has a higher energy per unit area than the material in the neighboring grains, and it can be shown that this energy is approximately constant.

2.9.4 Bulk defects

These are normally, (a) Casting defects, e.g. shrinkage cavities or gas holes.

(b) Working or forging defects, e.g. cracks, (c) Welding or joining defects, e.g. cracks, gas holes, slag penetration, corrosion regions.

The size of different types of structural defects are important when considering their effect and detection in materials, the relative sizes are shown in Fig. 2.30.

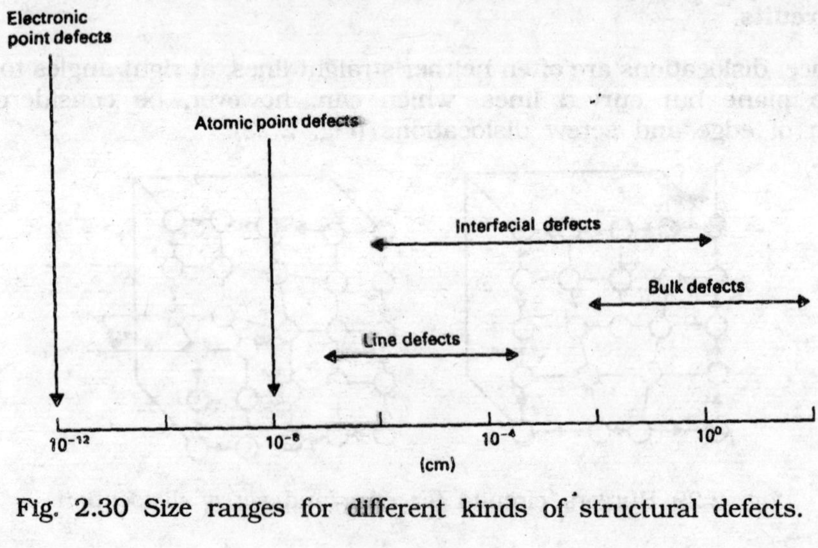

Fig. 2.30 Size ranges for different kinds of structural defects.

2.10 WORK HARDENING (STRAIN AGING OR STRAIN HARDENING)

Many metals possess a unique property in that, after undergoing some deformation, the metal possesses greater resistance to further plastic flow. As deformation proceeds whether by slip or by twinning the metal becomes harder and stronger and a stage is reached when further deformation is impossible. Any increase in stress will lead only to fracture. At this stage, when tensile strength and hardness are at a maximum and ductility is at minimum, the material is said to be work hardened.

The effect of plastic deformation briefly summarises an follows : If sufficient stress is applied to a metal, slip (or twinning) will take place in individual crystals. As deformation proceeds the capacity for further cold work diminishes as dislocations find themselves in positions from which further movement is impossible. An increase in stress is then necessary to initiate new dislocations and move these until they too become blocked. Ultimately a point is reached when no more mobile dislocations are available. This coincides with the point of maximum resistance to slip (the maximum strength and hardness) where no more deformation is possible and fracture will take place if an increase in stress is attempted. Further plastic deformation can only be carried out if the material is annealed.

During a cold worked process some 90% of the mechanical energy employed is converted to heat whilst the remaining 10% is retained in the material as potential energy in the form of elastically balanced strains in the lattice structure. The introduction of heat energy to the system during an annealing process permits sufficient mobility of the atoms so that dissipation of strain energy can take place. The initial relief of stress is followed by recrystallisation of the metal. The general effect of work hardening on strength, ductility, hardness and electrical resisvity are shown graphically in Fig. 2.31.

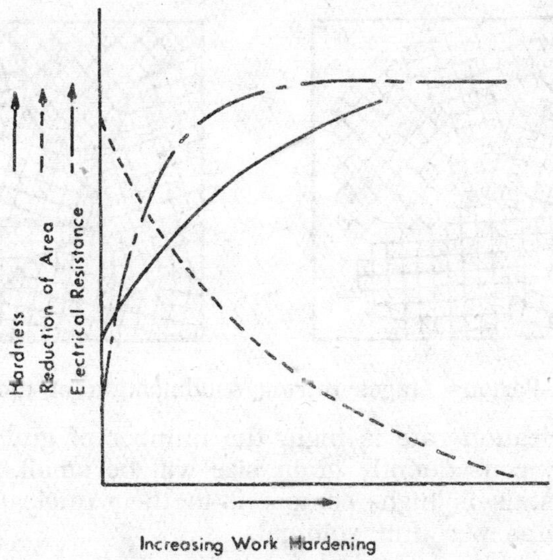

Fig. 2.31 Effect of work hardening on ductility, hardening and electrical resistance of metals.

2.11 GRAINS AND GRAIN BOUNDARIES

Metals commonly used for manufacturing various products are composed of many individual, randomly oriented crystals (grains). Now we are discussing polycrystals, that is many crystals.

When a mass of molten metal begins to solidify, crystals begin to form independently of each other, with random orientations and at various locations within the liquid mass (Fig. 2.32). Each of these crystals grows into a crystalline structure or grain. The number and size of the grains developed in a unit volume of the metal depends on the rate at which nucleation (initial stage of formation of crystals) takes place. The number of different sites in which individual crystals begin to form and the rate at which these crystals grow are important in the size of grains developed. **Generally, rapid cooling produces smaller grains, whereas slow cooling produces larger grains.**

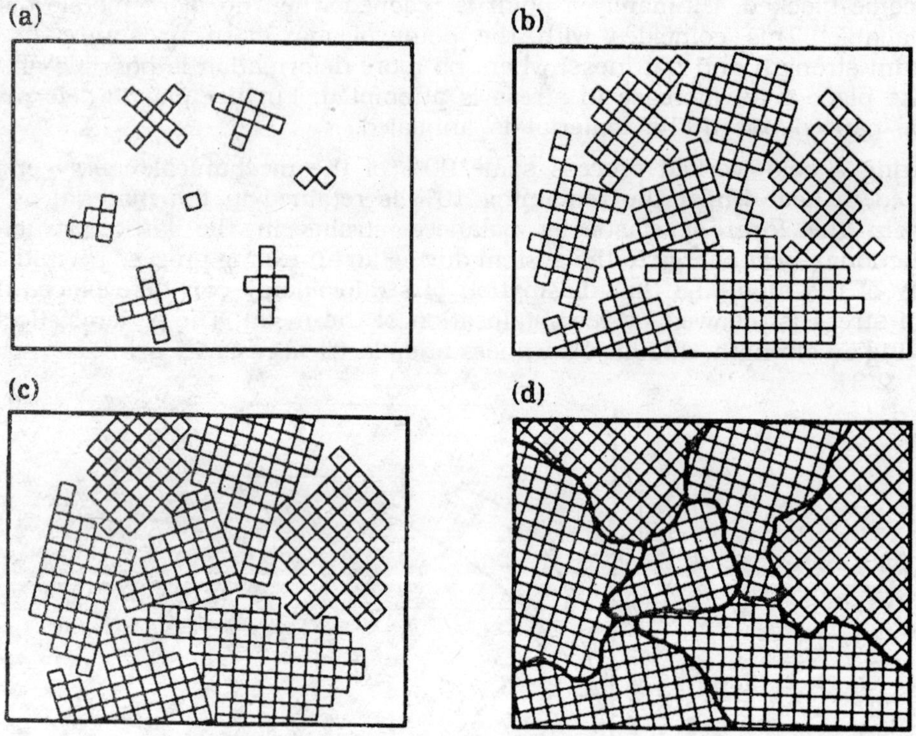

Fig. 2.32. Various stages during solidification of molten metals.

If the crystal nucleation rate is high, the number of grains in a unit volume of metal will be greater, consequently grain size will be small. Conversely, if the rate of growth of the crystals is high, compared to their nucleation rate, there will be fewer- but larger grains per unit volume.

The surfaces that separate these individual grains are called grain boundaries. It is to be noted that crystallographic orientation changes abruptly from one grain to the next across the grain boundaries. The behaviour of a single crystal or single grain is anisotropic (as a result of plastic deformation, the grains have elongated in one direction and contracted in other. Consequently, this piece of metal become anisotropic, and its properties in the vertical directions are different from those in the horizontal direction). The ideal behaviour of a piece of polycrystalline metal is isotropic because the grains have randon crystallographic orientations. Thus its properties do not vary with the direction of testing. In practice, however, this situation rarely exists.

2.12 RECOVERY, RECRYSTALLIZATION, AND GRAIN GROWTH

In plastic deformation at room temperature results in the deformation of grains and grain boundaries, a general increase in strength, and a decrease in ductility, as well as causing anisotropic behaviour. These effects can be reversed and the properties of the metal brought back to their original levels by heating it in a specific temperature range for a period of time. The temperature range and amount of time depend on the material and several other factors. Three events take place consequently during the heating process :

(1) Recovery

During recovery, which occurs at a certain temperature range below the recrystallization temperature of the metal, the stress in the highly deformed regions are relived and the number of mobile dislocations is reduced. Subgrain boundaries begin to form - called polygonization - with no appreciable change in mechanical properties, such as hardness and strength but some increase in ductility (Fig. 2.33).

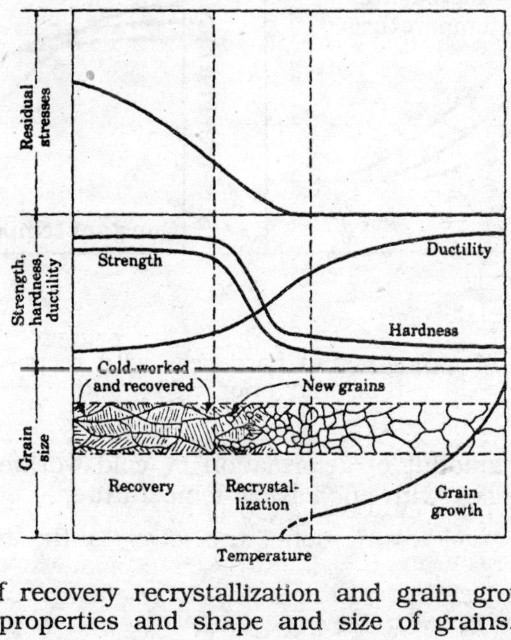

Fig. 2.33. Effect of recovery recrystallization and grain growth on mechanical properties and shape and size of grains.

(2) Recrystallization

The process in which, at a certain temperature range, new equiaxed and strain-free grains are formed, replacing the older grains, is called recrystallization. The temperature for recrystallization ranges approximately between 0.3 T_m and 0.5 T_m, Where T_m is the melting point of the metal on the absolute scale.

The recrystallization temperature is generally defined as the temperature at which complete recrystallization occurs within approximately one hour. Recrystallization decreases the density of dislocations and lowers the strength but raises the ductility of the metal (Fig. 2.33). Metals such as lead, tin, cadmium, and zinc recrystallize at about room temperature. Thus when these metals are cold worked, they do not work harden.

Recrystallization depends on the degree of prior cold work (work hardening); the higher the amount of cold work, the lower the temperature required for recrystallization to occur. the reason is that as the amount of cold work increases, the number of dislocation and the amount of energy stored in dislocations (stored energy) also increase. This energy supplies the work require for recrystallization. Recrystallization is a function of time because it involves, diffusion, that is, movement and exchange of atoms across grain boundaries.

The effects on recrystallization of temperature, time and reduction in the thickness or height of the workpiece by cold working are as follows (Fig. 2.34).

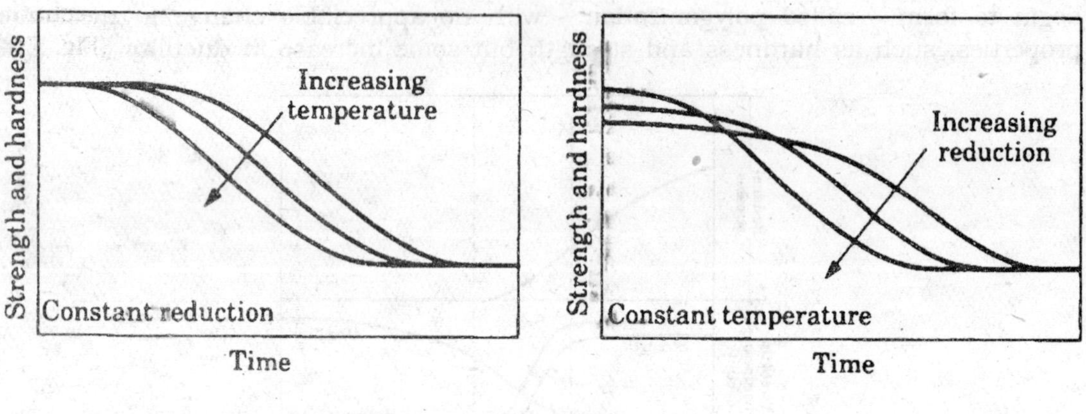

Fig. 2.34. Variation of strength and hardeness with recrystallization temperature, time and prior cold work.

(i) For a constant amount of deformation by cold working, the time required for recrystallization decreases with increasing temperature.

(ii) The more prior cold work done, the lower is the temperature required for recrystallization.

(iii) The higher the amount of deformation, the smaller the grain size becomes during recrystallization (Fig. 2.35). This is a common method of converting a coarse-grained structure to one of fine grain, with improved properties.

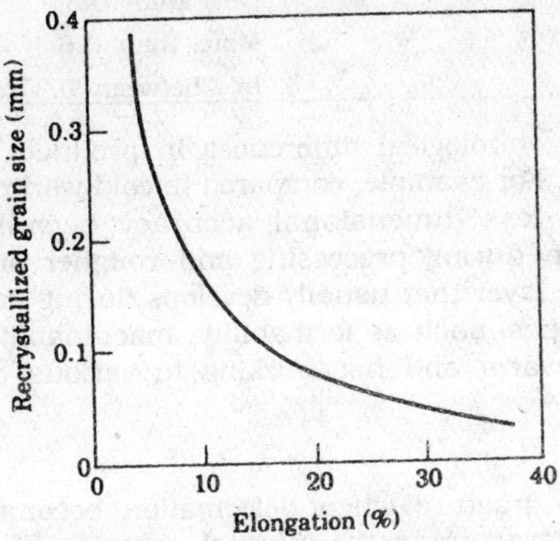

Fig. 2.35. The effect of prior cold work on the recrystallization grain size of alpha (α) brass. Below a critical elongation (Strain), typically 5%, no recrystallization occurs.

(iv) Anisotropy due to preferred orientation usually persists after recrystallization. To restore isotropy, a temperature higher than that required for recrystallization may be necessary.

(3) Grain growth

If we continue to raise temperature of the metal, the grains begin to grow and their size may eventually exceed the original grain size. This phenomenon is known as grain growth, and it has a slightly adverse effect on mechanical properties (Fig. 2.35). However, large grains produce a rough surface appearance on sheet metals called orange peel when they are stretched to form a part of when a piece of metal is subjected to compression, such as in forging operations.

2.13 HOMOLOGOUS TEMPERATURE RANGES FOR VARIOUS PROCESS

When plastic deformation is carried out at room temperature it is called **cold working**, and when carried out at above the recrystallization temperature, it is called **hot working**. As the name implies, **warm working** is carried out at higher than room temperature but lower than hot working temperature. This worm working is a compromise between cold and hot working. The temperature ranges for these three categories of plastic deformation are given in Table 2.4 in terms of a ratio, where T is the working temperature and T_m is the melting point of the metals both on the absolute scale. Although it is a dimensionless quantity this ratio is known as the homologous temperature.

Table 2.4. Homologous temperature ranges for various process.

SN	Process	T/Tm
1	Cold working	Less than 0.3
2	Hot working	More than 0.6
3	Warm working	In between 0.3 to 0.5

There are important technological differences in products that are processed by cold, warm or hot working. For example, compared to cold worked products, hot worked products generally have less dimensional accuracy because of uneven thermal expansion and contraction during processing and rougher surface appearance and finish because of the oxide layer that usually develops during heating. Other important manufacturing characteristics, such as formability, machinability, and weldability, are also affected by cold, warm and hot working to various degrees.

2.14 FRACTURE

A material will fail by fracture when deformation becomes excessive and work hardening effects reach a maximum; the internal cohesion of the material is broken down, this resulting in the separation of one section from another.

Fractures are usually said to be either **ductile or brittle**, depending upon the amount of plastic deformation preceding failure.

When a ductile material has a gradually increasing stress applied to it behaves elastically up to a limiting stress and then beyond that stress plastic deformation occurs. In the case of a tensile stress this deformation takes the form of necking (Fig. 2.36a). As the stress is increased, the cross-sectional area of the material becomes considerably reduced until at some stress failure occurs. The fracture shows a typical corse and cup formation, which results because under the action of the increasing stress small internal cracks form which gradually grow in size until there is an internal, almost horizontal, crack. The final fracture occurs when the material shears at an angle of 45° to the axis of the direct stress. This type of failure is known as a **ductile fracture**. A characteristic of a ductile fracture is that the surfaces of the fractured material are dull or fibrous.

Materials can fail in a ductile manner in compression. Such failures show a characteristic bulge and a series of axial cracks around the edge of the material (Fig. 2.36b).

Another form of failure is known as brittle fracture. Fig. 2.37a shows the possible forms of a brittle tensile failure with metals. The surfaces of the fractured material appear bright and granular due to the reflection of light from individual crystal surfaces. This is because the fracture has taken place by the separation of grains within the material along specific crystal planes of atoms. Such planes are called **cleavage planes**.

Materials can fail in a brittle manner in compression. Fig. 2.37b shows compressive failure.

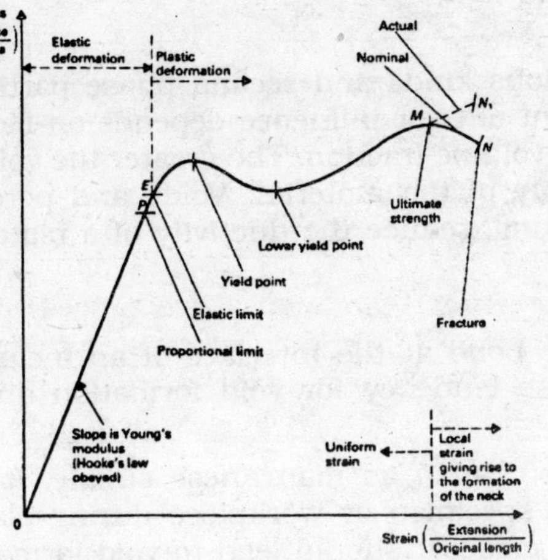

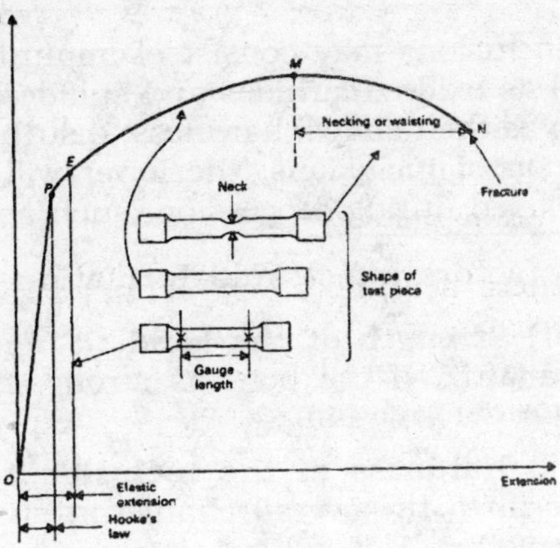

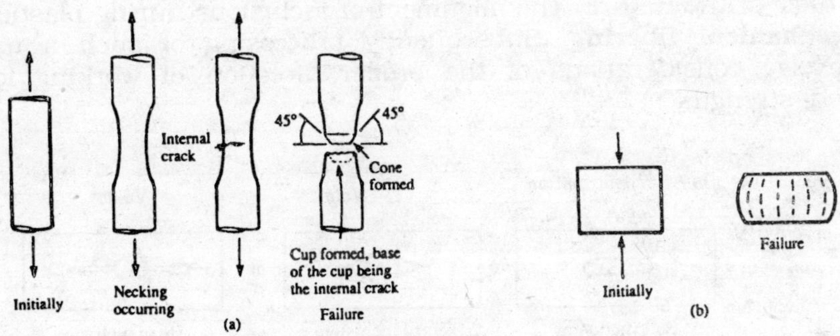

Fig. 2.36. Ductile failure (a) Tensile, (b) Compressive.

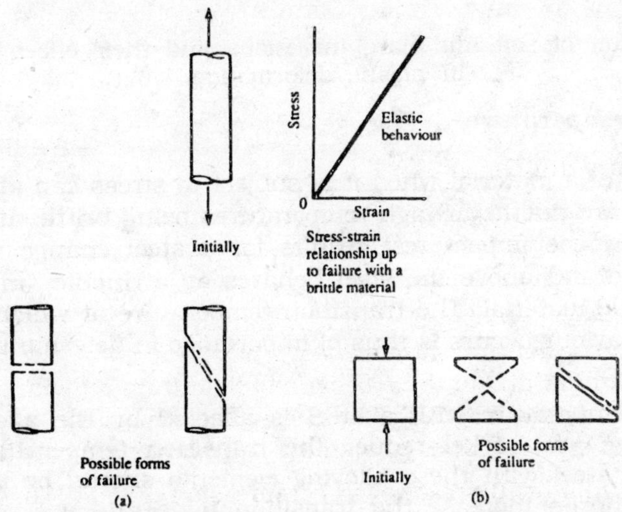

Fig. 2.37. Brittle failure (a) Tensile, (b) Compressive.

2.14.1 Effects of inclusions

Inclusions may consist of impurities of various kinds and second phase particles, such as oxides, carbides and sulfides. The extent of their influence depends on factors such as their shape, hardness, distribution and volume fraction. The greater the volume fraction of inclusions, the lower will be ductility of the material. Voids and porosity developed during processing, such as from casting, reduce the ductivity of a material.

Two factors affect void formation

(1) Strength of the bond : Strength of the bond at the interface of an inclusion and matrix. If the bond is strong, there is less tendency for void formation during plastic deformation.

(2) Hardness of the inclusion : It it is soft, such as manganese sulfide, it will conform to the overall change in shape of the specimen or workpiece during plastic deformation. If it is hard, such as a carbide or oxide, it should lead to void formation (Fig. 2.38). Hard inclusions may also break up into smaller particles during deformation because of their brittle nature. The alignment of inclusions during plastic deformation leads to **mechanical fibering.** Subsequently processing of such a material must therefore involve considerations of the proper direction of working for maximum ductivity and strength.

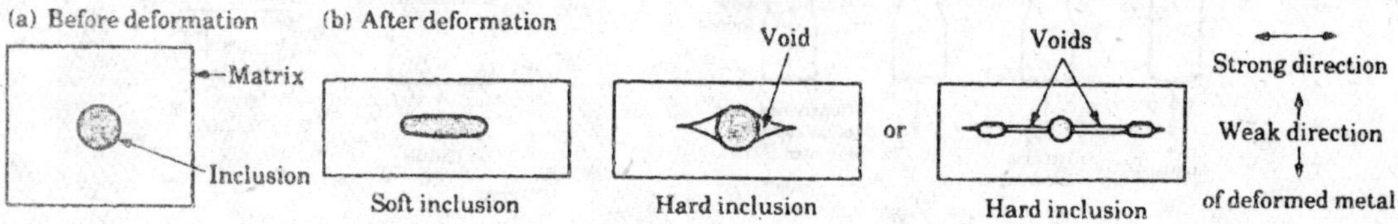

Fig. 2.38. Deformation of soft and hard inclusions and their effect on void formation in plastic deformation.

2.14.2. Transition temperature

The temperature of a material when it is subject to stress can affect its behaviour, many metals which are ductile at high temperatures being brittle at low temperature. Fig. 2.39 shows how the impact test results for a steel change with temperature. At room temperature and above the steel behaves as a ductile material; below 0°C it behaves as a brittle material. The transition temperature at which the change from ductile to brittle behaviour occurs is thus of importance in determining how a material will behave in service.

The transition temperature with a steel is affected by the alloying elements in the steel. Manganese and nickel reduce the transition temperature. Thus for low temperature work a steel with these alloying elements should be preferred. Carbon, nitrogen and phosphorus increase the transition temperature.

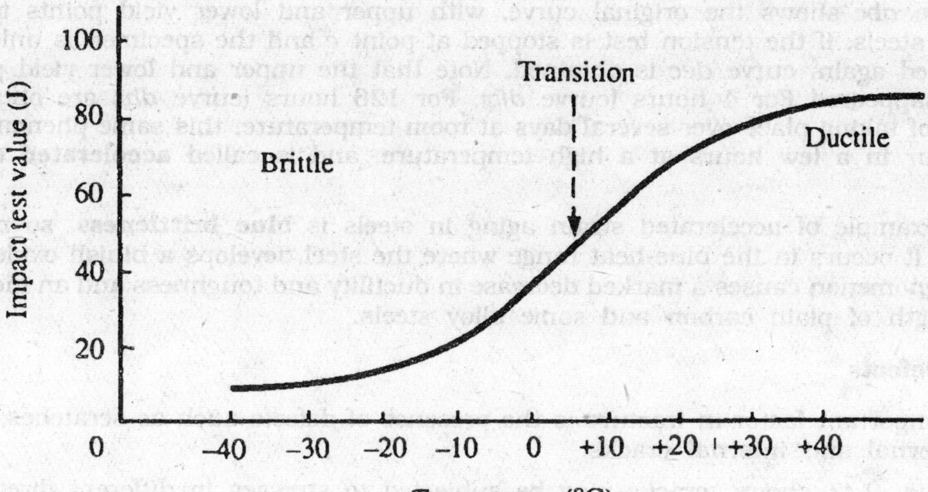

Fig. 2.39. Transition temperature.

2.14.3 Strain aging

Strain aging is a phenomenon in which carbon atoms in steels segregate to dislocations, thereby pinning them and thus increasing the resistance to dislocation movement. The result is increased strength and reduced ductility. The effects of strain aging on the shape of the stress strain curve for a low carbon steel at room temperature are shown in Fig. 2.40.

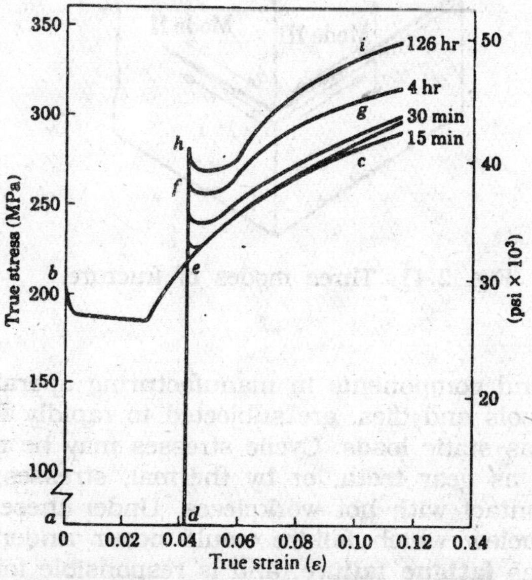

Fig. 2.40. Strain aging and its effect.

Curve *abc* shows the original curve, with upper and lower yield points typical of these steels. If the tension test is stopped at point *e* and the specimen is unloaded and tested again, curve *dec* is obtained. Note that the upper and lower yield points have disappeared For 4 hours (curve *dfg*), For 126 hours (curve *dhi*) are obtained. Instead of taking place over several days at room temperature, this same phenomenon can occur in a few hours at a high temperature, and is called **accelerated strain aging**.

An example of accelerated strain aging in steels is **blue brittleness**, so named because it occurs in the blue-heat range where the steel develops a bluish oxide film. This phenomenon causes a marked decrease in ductility and toughness and an increase in strength of plain carbon and some alloy steels.

2.14.4 Defects

An important factor in fracture is the presence of defects such as scratches, flows and external and internal cracks.

As Fig. 2.41 shows, cracks may be subjected to stresses in different directions. Mode I is tensile stress applied perpendicular to the crack. Modes II and III are shear stresses applied in two different directions. Tearing paper, cutting sheet metal with shears or opening flip-top cans are examples of mode III fracture. Mode II is rare. mode I has been studied extensively in engineering structure and components.

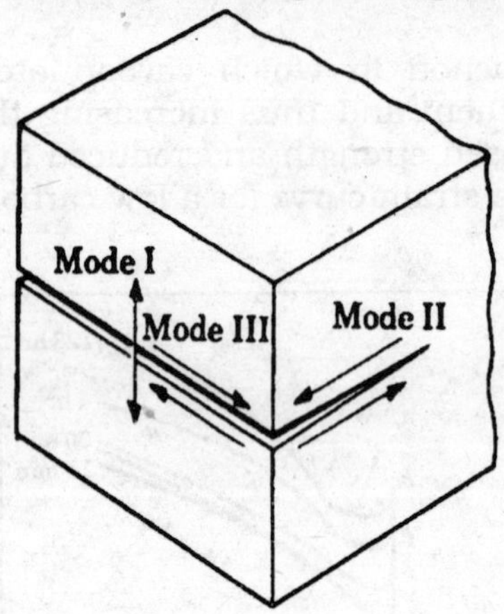

Fig. 2.41. Three modes of fracture.

2.15 FATIGUE

Various structures and components in manufacturing operations, such as cams, shafts, springs, gears, tools and dies, are subjected to rapidly fluctuating (cycling or periodic) loads, as well as static loads. Cyclic stresses may be caused by fluctuating mechanical loads, such as gear teeth, or by thermal, stresses, such as a cool die coming into repeated contact with hot workpieces. Under these conditions, the part fails at a stress level below which failure would occur under static loading. This phenomenon is known as **fatigue failure**, and is responsible for at least 80% of the failures in modern engineering components.

2.15.1 Fatigue failure with a metal

A fatigue crack often starts at some point of stress concentration. This point of origin of the failure can be seen on the failed material as a smooth, flat, semicircular as a smooth, flat, semicircular or elliptical region, often referred to as the nucleus. Surrounding the nucleus is a burnished zone with ribbed markings. This smooth zone is produced by the crack propogating relatively slowly through the material and the resulting fractured surfaces rubbing together during the alternating stressing of the component. When the component has become so weakened by the crack that it is no longer able to carry the load, the final abrupt fracture occurs, which shows a typically crystalline appearance. Fig. 2.42 shows the various stages in the growth of a fatigue crack failure.

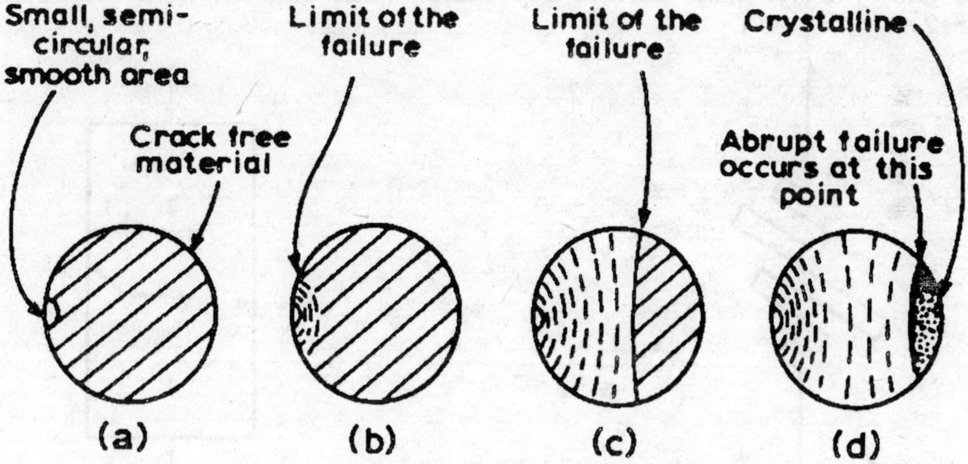

Fig. 2.42. Fatigue failure with a metal
(a) The nucleus, limit of the failure, (b) The crack grows slowly (c) The crack continuing to grow slowly (d) Complete failure.

2.15.2 Stages in fatigue failure

A metal progressing to fatigue failure can be considered to pass through three stages

Stage 1 - The nucleus and initial crack : This is the initial zone of the fatigue crack. Under repeated stressing slip occurs as a movement of dislocations. Slip may occur one way on one slip plane and the reverse way on an adjacent slip plane during the reverse stress cycle. The reverse direction of slip on the first slip plane is inhibited by local work hardening, i.e. an increase in dislocation density in this slip plane as a result of its plastic deformation. The result of such behaviour is a number of slip bands, essentially a series of grooves and extended tongues of metal from the metal surface (Fig. 2.43). From this discontinuity the fatigue cracks start. The initial crack growth is in a plane which is about 45° to the direction of the applied stress. This is a consequence of the slip occuring as a result of shear.

After a few grains the direction of the crack changes to be at right angles to the applied stress direction. This change in direction in the beginning of stage 2.

Stage 2 - Crack propogation : The mode of failure of the material is stage 1 is a shear mode, in stage 2 the mode is a tensile mode with the crack being forced open (Fig. 2.44). The stage 2 crack surfaces show a series of fine striations. This striations are a series of ridges and result from the alternating stress applied to the material, each stress cycle producing a striation. The striations are concentric on the nucleus and the spacing between successive striations increases as the crack propogates out from the nucleus.

Stage 3 - Failure : The stage 2 crack continues growing until a point is reached when there is a sudden complete failure of the material. This final fracture zone is likely to show a crystalline surface appearance, quite different from the striations of stage 2.

Fig. 2.43. Slip bands at the nucleus of fatigue failure.

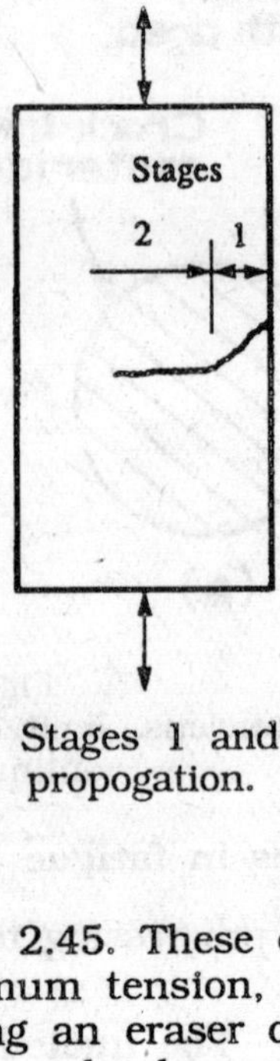

Fig. 2.44. Stages 1 and 2 cracke propogation.

2.15.3 S-N curves

A typical plot, known as S-N curves, is shown in Fig. 2.45. These curves are based on complete reversal of the stress- that is, maximum tension, maximum compression, and so on, such as that obtained by bending an eraser or piece of wire alternately in one direction, then the other. The test can also be performed on a rotating shaft with a constant downward load.

Endurance strength (also called fatigue strength) is the maximum, completely reversed stress under which a material will fail after it has experienced the stress or a specified number of cycles. Thus, when a value for the endurance strength of a material is stated, it must be accompanied by the number of stress cycle.

The endurance limit (also called fatigue limit) is the maximum, completely reversed stress for which it is assumed a material will "never" fail regardless of the number of stress cycles. Thus, the endurance limit is stated without an associated number of cycles to failure. It is generally accepted as being standard that

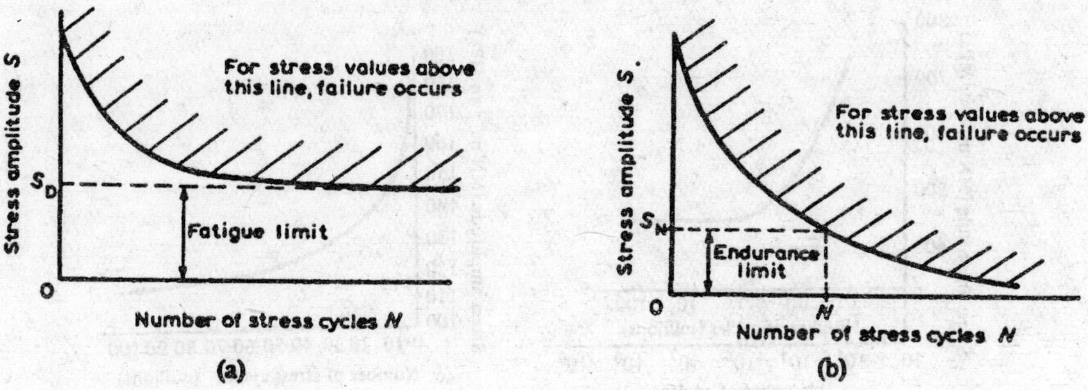

Fig. 2.45. Typical S-N curves for (a) A steel (b) A non ferrous alloy.

ferrous materials that survive at least 10^7 cycles of stress reversals will have an infinite life. Non ferrous materials such as brass, copper and copper alloys, aluminum, and magnesium do not have an endurance limit; therefore, only an **endurance strength** can be established for these materials.

The **endurance limit** for metals is related to their ultimate tensile strength (Fig. 2.46). For steels, the endurance limit is about one half their tensile strength. Although many metals, especially steels, have a definite endurance limit, aluminium alloys do not have one, and the S-N curve, continues its downward trend (Fig. 2.47(a) and Fig. 2.47(b).

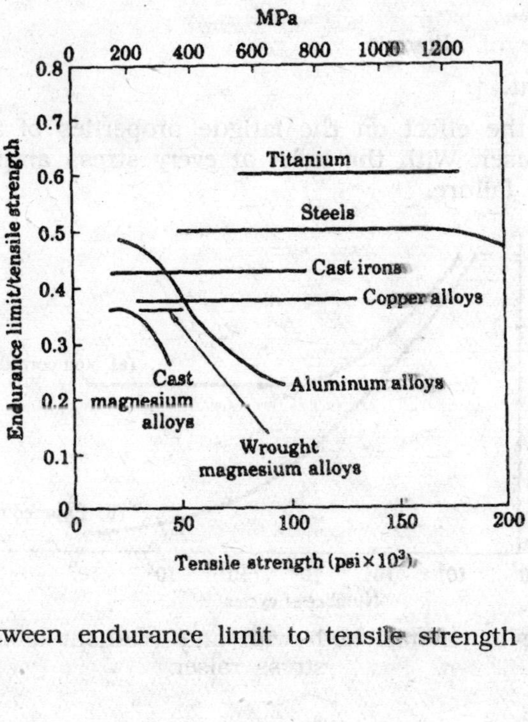

Fig. 2.46. Graph between endurance limit to tensile strength for various metals.

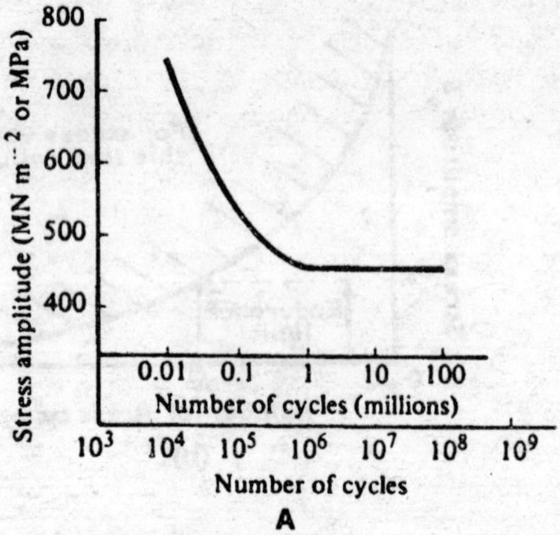

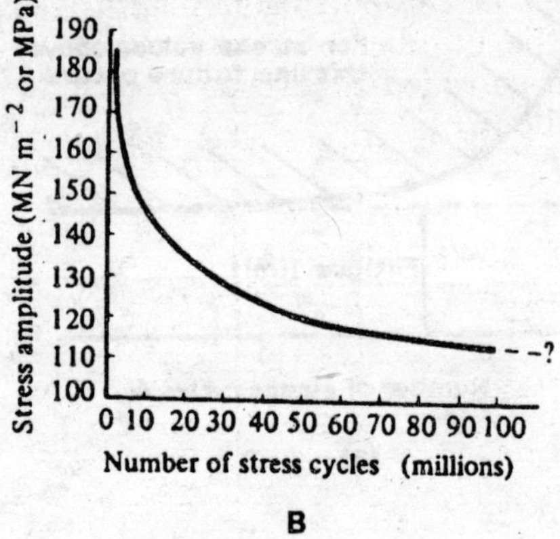

Fig. 2.47. Experimental values of S-N curves (a) for steel (b) for an aluminium alloy

2.15.4 Factors affecting the fatigue properties of metals

The main factors affecting the fatigue properties of a component are :

(1) Stress concentrations caused by component design.

(2) Corrosion.

(3) Residual stresses.

(4) Surface finish/treatment

(5) Temperature.

(6) Microstructure of alloy.

(7) Heat treatment.

Fig. 2.48 shows the effect on the fatigue properties of a steel of a small hole acting as a stress raiser. With the hole, at every stress amplitude value less cycles are needed to reach failure.

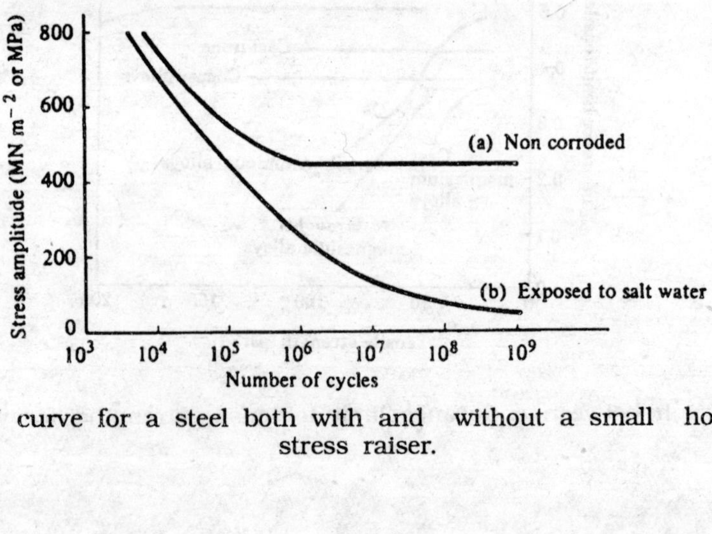

Fig. 2.48. S-N curve for a steel both with and without a small hole acting as a stress raiser.

Fig. 2.49 shows the effect on the properties of a steel of exposure to salt solution. The effect of the corrosion resulting from the salt solution attack on the steel is to reduce the number of stress cycles need to reach failure for every stress amplitude. The non-corroded steel has a fatigue limit of 450 MPa, the corroded steel has no fatigue limit.

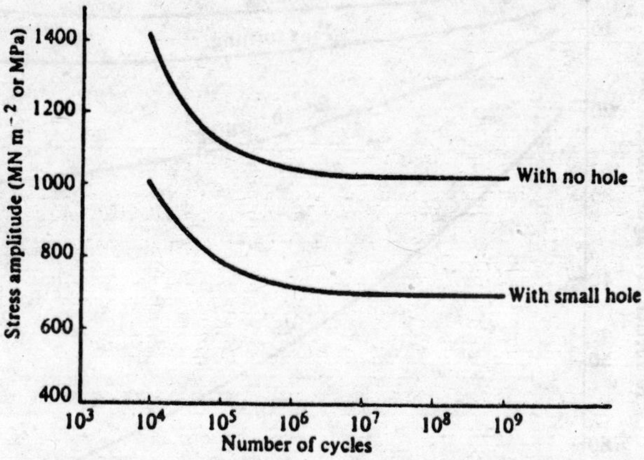

Fig. 2.49. S-N curve for a steel (a) with no corrosion (b) corroded to exposure to salt solution.

Fig. 2.50 shows the effect of carburizing a hardened steel. Many machining processes result in the production of surface tensile residual stresses and so result in poorer fatigue properties.

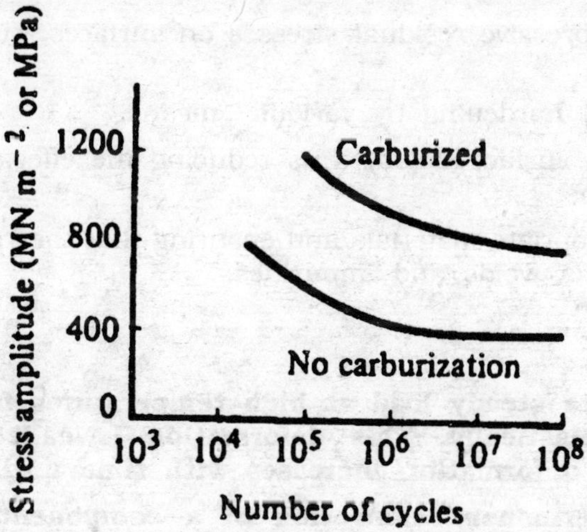

Fig. 2.50. S-N curve for a steel, showing effect of carburization.

Fatigue life is greatly influenced by the method of preparation of the surfaces of the part of specimen (Fig. 2.51). The fatigue strength of manufactured products can be improved generally by the following methods :

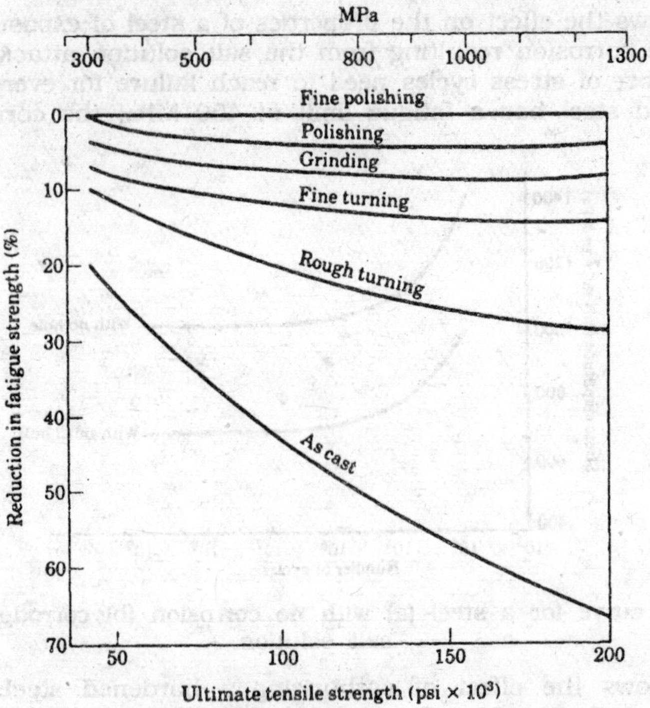

Fig. 2.51. Reduction in fatigue strength of cast steels subjected to various surface finishing operations.

(1) Inducing compressive residual stresses on surfaces, such as by shot peening or roller burnishing.

(2) Surface (case) hardening by various means.

(3) Providing fine surface finish, thus reducing the effects of notches and other surfaces imperfections.

(4) Selecting appropriate materials and ensuring that they are free from significant amounts of inclusions, voids and impurities.

2.16 CREEP

Materials under a steady load at high temperatures for a long time period begin to deform plastically. This deformation is called creep and is time dependent because deformation increases with time until rupture takes place.

Creep is the permanent elongation of a component under a static load maintained for a period of time. It is a phenomenon of metals and certain nonmetallic materials, such as thermoplastics and rubbers, and can occur at any temperature.

Fatigue failure would most likely take place early in the life of a material. However, because steam or gas turbines etc., are designed to operate at elevated temperatures for long periods of time there is great interest in the phenomenon of creep.

For metals and their alloys, creep of any significance occurs at elevated temperatures, beginning at about 200°C for aluminium alloys, and up to about 1500°C for refractory alloys. The mechanism of creep at elevated temperature in metals is generally attributed to grain boundary sliding. Creep is especially important in high temperature applications, such as gas turbine blades and similar components in jet engines and rocket motors. High pressure steam lines and nuclear fuel elements also are subject to creep. Creep deformation also can occur in tools and dies that are subjected to high stresses at elevated temperature during hot working operations such as forging and extrusion.

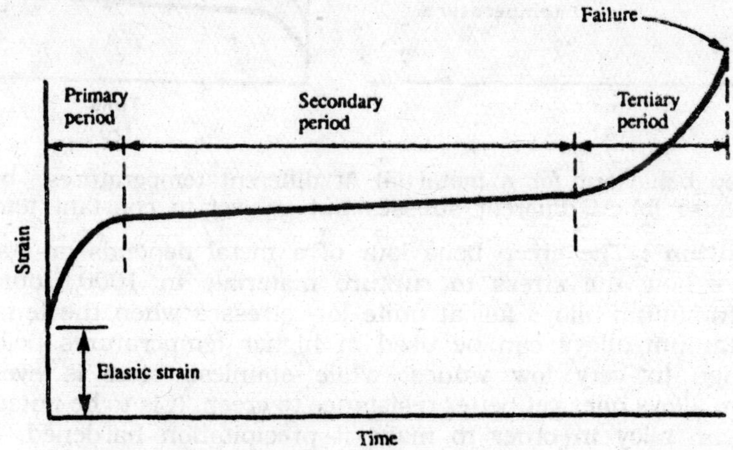

Fig. 2.52. Typical creep curve for a metal.

Fig. 2.52 shows the general form of results from a creep test. Following an 'instantaneous' elastic strain region the curve generally has three parts.

The initial strain takes place, almost immediately and consists of the elastic strain plus plastic strain if the deformation extends beyond the yield point. The **primary creep (first stage)** shows a decreasing rate of elongation due to strain hardening. **The secondary creep (second stage)** begins at a minimum strain rate and remain constant because of the balancing effects of strain hardening and annealing. **The teritary creep (third stage)** shows a rapid increase of the creep rate accompanied by severe necking and ultimate rupture.

A family of graphs can be produced showing the creep for different initial stresses at a particular temperature and for different temperatures for a particular initial stress (Fig. 2.53).

2.16.1 Factors affecting creep with metals

The main factors affecting creep behaviour with metals are :

(1) Temperature : The higher the temperature the greater the creep at a particular stress (Fig. 2.53).

(2) Stress : The higher the stress at a particular temperature the greater the creep (Fig. 2.53).

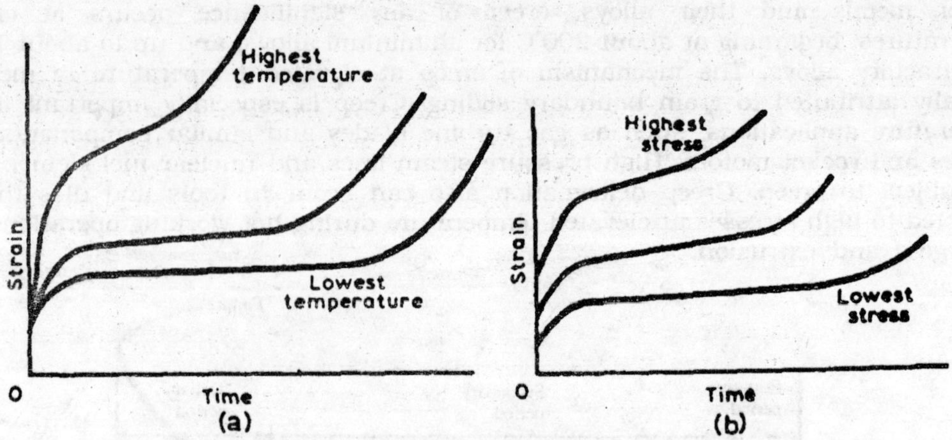

Fig. 2.53. Creep behaviour for a metal (a) At different temperatures but subject to constant stress (b) At different stresses but subject to constant temperature.

(3) Composition : The creep behaviour of a metal depends on its composition. Fig. 2.54 shows how the stress to rupture materials in 1000 hours varies with temperature. Aluminium alloys fail at quite low stresses when the temperature rises over 200°C. Titanium alloys can be used at higher temperatures before the stress to rupture drops to very low values, while stainless steel is even better and nickel-chromium alloys offer yet better resistance to creep. It is to be noted that addition of elements to an alloy in order to make it precipitation hardened.

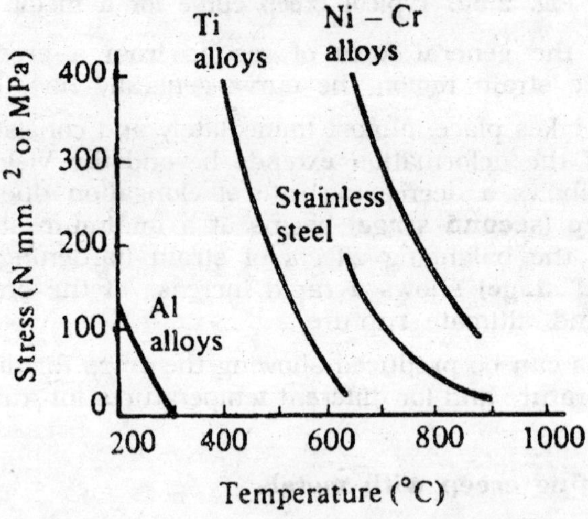

Fig. 2.54. Stress to rupture in 1000 hours for different materials.

(4) Grain Size : Grain size can have an effect on the creep behaviour of a material. A low temperatures high grained material has the best creep properties while at high temperature coarse grained material is best.

2.17 HARDNESS

The hardness of a material can be defined in terms of the ability of the material to resist plastic scratching or indentation. There is no absolute hardness scale and hardness values are expressed generally in terms of a scale associated with a particular test. In the Brinell hardness test, a hardened steel ball is passed into the surface of the material and the spherical area of the indentation (Fig. 2.55) determined for the load used. Taking into account the magnitude of the force used, this gives a Brinell hardness number (HB). The larger this number the harder the material.

A particle relationship has been observed between the ultimate tensile strength (UTS) and Brinell hardness number (HB) for steels

UTS = 3.5 (HB), where UTS is in MPa.

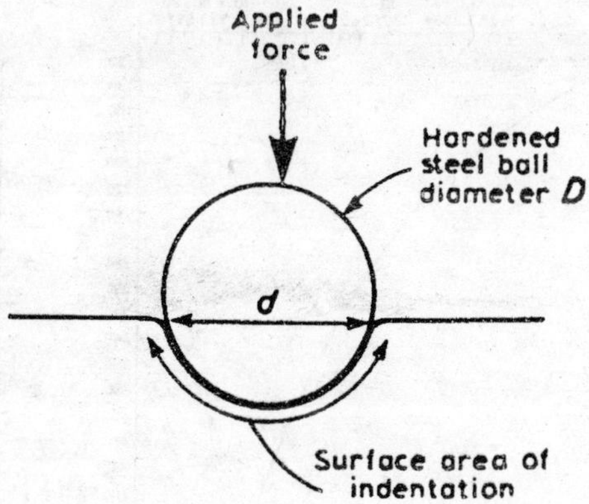

Fig. 2.55. The Brinell hardness test.

The Vickers hardness test the indenter used is specially shaped diamond and again the area of the indentation is measured. The Vickers hardness number (HV) is obtained from the area and the force. In the Rockwell hardness test, the indenter is a standard diamond corse or hardened steel ball and the permanent increase in depth of penetration as a result of change in the load applied to the indenter is measured. The Rockwell hardness (HR) is given by (E-e), where E is a constant determined by the type of indenter used and e is the depth of penetration. The Shore Scleroscope hardness test measures the dynamic hardness, not indentation hardness as the others do. Measurement is made of the height of rebound of the indenter after its fall from a fixed height. The Knoop test is a microhardness test because of the light loado utilized; hence it is suitable for very small or thin specimens and for brittle materials, such as gemstones, carbides, and glass. This test is also used in measuring the hardness of individual grains in a metal.

Commonly used hardness numbers are not simple convertible each into the other. Fig. 2.56 shows a conversion plot between Brinell, Rockwell, and Vickers hardness numbers. Also, Fig. 2.57 shows the conversion plot between Brinell, Rockwell, Vickers and shore hardness numbers, and the ultimate tensile strength of steel. It is to be noted that these curves are nonlinear and that the related values are only approximate.

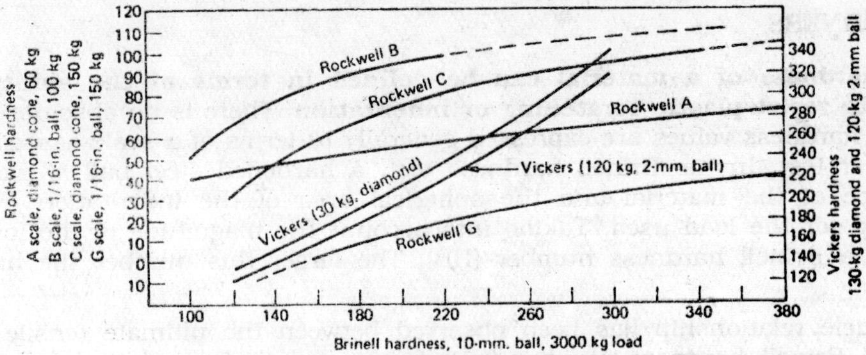

Fig. 2.56. Conversion chart

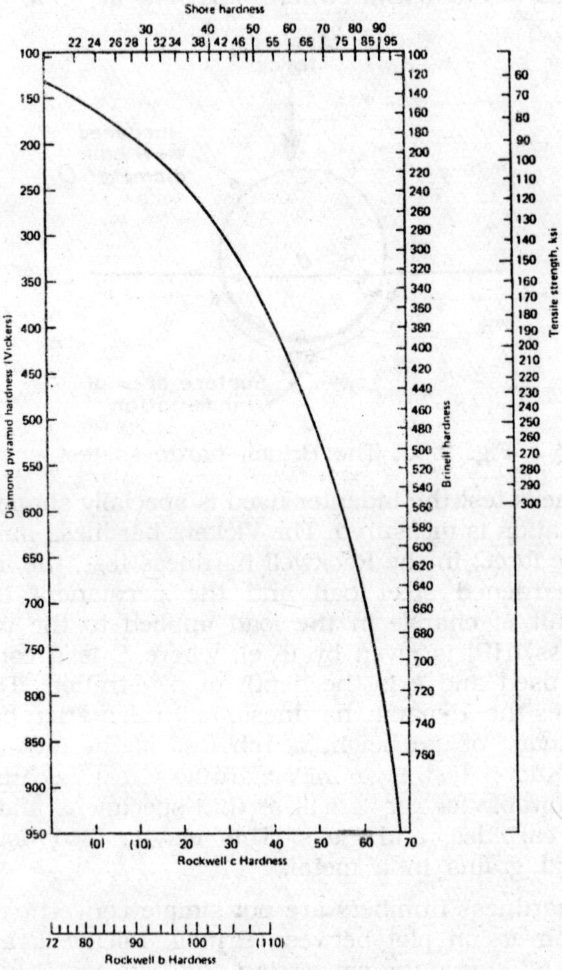

Fig. 2.57. Hardness conversion and relationship to the ultimate tensile stress of test.

2.18 IMPACT

A brittle material absorbs little energy in breaking but a ductile material absorbs a great amount. **The term toughness is often used to describe the ability of a material to withstand shock loads, the tougher the material the more it is able to withstand such loads without breaking.** The impact tests can be considered as an indicator of toughness, the greater the value given by the Izod or Charpy test the tougher the material. The data from one test cannot be directly compared with that from the other test, hence the need to specify the test used.

One use of impact tests is to determine whether heat treatment of metals has been successfully carried out. A comparatively small change in heat treatment can lead to quite noticeable changes in impact test results. The changes can be considerably more pronounced than changes in other mechanical properties, e.g. percentage elongation or tensile strength. Fig. 2.58 shows the effect on the Izod impact test results for cold worked mild steel annealed to different temperatures. The use of an impact test could then indicate whether annealing has been carried out to the corred temperature.

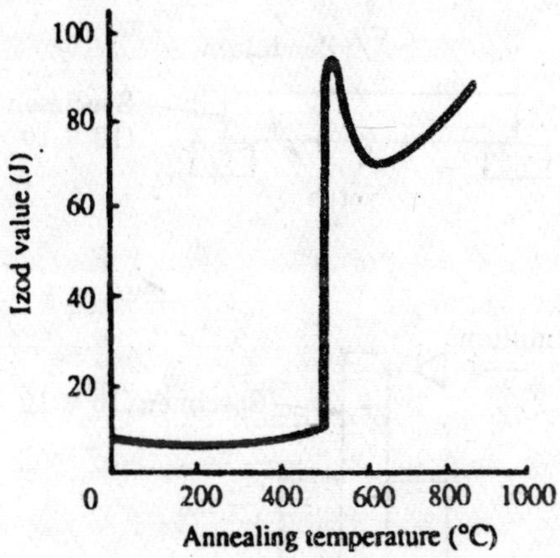

Fig. 2.58. Effect of annealing temperature on Izod test values.

The properties of metals change with temperature. For example, a 0.2% carbon steel undergoes a gradual transition from ductile to a brittle material at a temperature of about room temperature (Fig. 2.59). At about -25°C the material is a brittle material with a charpy V-notch impact energy of only about 4J, whereas at about 100°C it is ductile with an impact energy of about 120J. This type of change from ductile to a brittle material can be charted by impact test results and the behaviour of the material at the various temperatures predicted.

A typical impact test consists of placing a notched specimen in an impact tester and breaking it with a swinging pendulum. In the charpy test the specimen is supported at both ends (Fig. 2.60a), whereas in the Izod test it is supported at one end like a cantilever beam (Fig. 2.60b). From the amount of swinging of the pendulum, the energy dissipated in breaking the specimen is obtained. This energy is the impact toughness of the material.

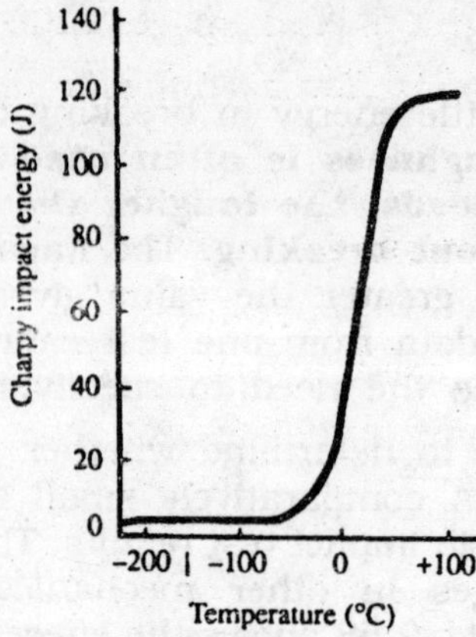

Fig. 2.59. Effect of temperature on the charpy V-notch impact energies for 0.2% carbon steel

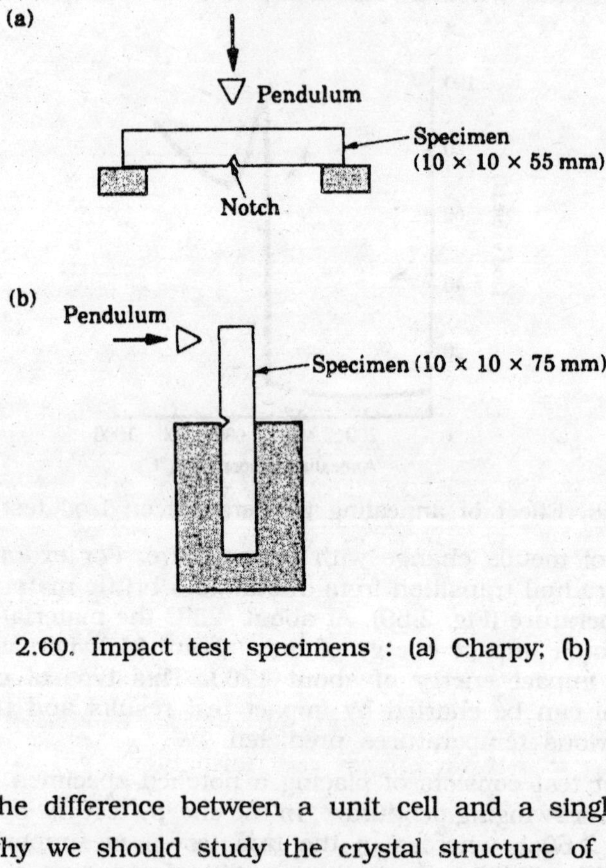

Fig. 2.60. Impact test specimens : (a) Charpy; (b) Izod.

EXERCISES

A : Theory

2.1 What is the difference between a unit cell and a single crystal?

2.2 Explain why we should study the crystal structure of metals.

2.3 What effects does recrystallization have on the properties of metals?

2.4 What is the significance of a slip system?

2.5 Explain the terms unit cell and space lattice.

2.6 How does a crystalline material differ in structure from an amorphous material?

2.7 Distinguish between elastic and plastic deformation of metals and give a simple explanation of such deformation in terms of the block slip model.

2.8 Explain why slip only occurs in particular directions in a crystal.

2.9 Explain the nature of point and line defects in crystals.

2.10 Explain how dislocations density, grain size and foreign atoms affect the mobility of dislocations within a metal and hence the yield stress of that metal.

2.11 Explain in terms of dislocations (a) work hardening (b) recovery and (c) annealing.

2.12 Explain twinning in tension.

2.13 Explain strain twins and annealing twins.

2.14 Draw a neat sketch showing relative atom movements in slip and twinning.

2.15 Draw neat sketches of an edge dislocation and a screw dislocations and explain them.

2.16 Describe crystal defects.

2.17 Explain density of dislocations.

2.18 Explain work hardening or strain hardening.

2.19 Draw neat sketches showing the effects of hardening on ductility, hardness and electric resistance of metals.

2.20 Explain various stages during solidification of molten metals.

2.21 Explain the advantages and limitations of cold, hot and warm working of metals.

2.22 What is the significance of metals such as lead and tin having recrystallization temperatures at about room temperature?

2.23 Explain why the strength of a polycrystalline metal at room.

2.24 Describe the technique by which you would reduce the orange peel effect.

2.25 Distinguish between the forms of ductile and brittle fractures.

2.26 Describe the nucleation and growth of a ductile fracture.

2.27 Explain the effects of inclusions on ductile materials.

2.28 Explain the two factors affect void formation.

2.29 Explain the effects of transition temperature which are ductile.

2.30 Explain the transition temperature with steel is affected by the alloying elements in the steel.

2.31 Explain the effects of strain aging.

2.32 Explain the three modes of fracture.

2.33 Explain the important factor in fracture is the presence of defects.

2.34 Explain what is meant by fatigue failure.

2.35 Explain the terms fatigue limit and endurance limit.

2.36 Explain S-N graph and state the information that can be extracted from the graph.

2.37 Explain stages in fatigue failure.

2.38 Draw S-N curves for steel and an aluminium.

2.39 Describe factors affecting the fatigue properties of metals.

2.40 Explain what is meant by creep.

2.41 Describe the effect of increased stress and increased temperature on the creep behaviour of materials.

2.42 Describe the creep mechanisms that occur with a metal in the primary, secondary and tertiary stages of creep.

2.43 Describe factors affecting creep with metals.

2.44 Define hardness

2.45 Define toughness

2.46 Draw a conversion chart between different hardness numbers.

2.47 Draw hardness conversion and relationship to the ultimate tensile stress for test chart.

2.48 Draw a graph showing the effect of annealing temperatures on Izod test values.

2.49 Draw a graph showing the effect on the charpy V-notch impact energies for 0.2% carbon steel.

3

ENGINEERING MATERIALS

3.1 INTRODUCTION

All materials, whether solid, liquid or gaseous, can be classified as elements, compound or mixtures on the basis of chemical composition and internal structure.

Elements are simple substances which cannot be broken down into chemically simple substances. Of the known elements only a very small number have importance as engineering materials in their elemental (combined) states.

Compounds are formed when two or more elements combine chemically in fixed proportions by weight; all compounds can be broken down, by one means or another, into their component elements. Many important engineering materials exist in the form of chemical compounds.

Mixtures are formed when two or more pure substances (elements or compounds) are mechanically mixed together in any proportion; thus, a mixture differs from a compound in that it is, in theory anyway, mechanically or physically separable.

3.2 METALS AND NON-METALS

The elements may be roughly divided according to their properties into two groups, the metals and non metals.

The engineer looks at metals from a very different viewpoint from that of the pure scientist, being more concerned with their properties and applications that with their behaviour in chemical reactions.

3.3 Metals

A typical classification of metals are used in engineering would be as follows : (1) Pure, (2) Alloyed (3) Coated, (4) Clad (5) Ferrous (6) Non Ferrous and (7) Sintered.

3.3.1 Pure Metals

Pure metals are those obtained when an ore is refined to yield a metallic element. In practice it is very difficult to obtain a perfectly pure metals, and indeed many metals are worthless to the engineer when in this state, but by using specialised and, incidentally, often very expensive techniques some metals can be obtained 99.999% pure. Copper and aluminium are common examples of this.

3.3.2 Alloy

An alloy is formed when two or more relatively pure metals are melted together to form a new metal having in many instance, properties quite different from those of the two metals used in its manufacture. An outstanding example of this is the alloy formed when 18% chromium and 8% nickel are added to low carbon steel, forming a steel which will not rust or corrode, this being quite different from the behaviour of the original low carbon steel.

3.3.3 Coated metals

Coated metals, including such metals as tinplate, terneplate, galvanised iron, and zincanneal, are manufactured so that the desirable properties of both metals can be "married together". For instance, tinplate (mild steel sheet coated with a very thin layer of tin) enables steel to be used for food containers, while galvanised iron (mild steel coated with a thin layer of zinc) enables strong this sheets of steel to be used for constructional purposes without fear of rusting.

3.3.4 Clad metals

Clad metals are also made to take advantage of the properties of both materials used, but in this case a "sandwich" is made of the metals concerned. For instances duralumin is clad with thin sheets of pure aluminium for use an exposed surfaces, since the surface layers of aluminium resist corrosion while the center layer of duralumin provides high strength. Stainless steel is often backed by a relatively thick layer of mild steel by rolling the two metals together while they are red hot. This gives a clad metal which will not rust or corrode on one surface, and which is still relatively inexpensive to manufacture. A typical application of such a clad metal in the construction of oil storage tanks, where rust and corrosion cannot be tolerated inside the tank, but where the cost of fabrication using thick sheets of stainless steel would be prohibitive.

3.3.5 Ferrous metals

A ferrous metal is one composed principally of iron, while non- ferrous metals are those composed principally of metals other than iron. It should be noted from the outset that a non ferrous alloy may contain iron in same small proportion (for instance, high-tensile brass often contains about 1% iron by weight) and that ferrous alloys may contain quite significant proportions of non ferrous metals.

Of the vast number of non-ferrous metals available to man, only seven are available to a low enough cost and in sufficient quantity to serve as bases for common engineering metals. Three of these (copper, lead and tin) have been used for thousands of years, while the remaining four (aluminium, magnesium, nickel and zinc) are the more recent additions, achieving commercial importance only in the latter part of the nineteenth century. At least fourteen other metals - antimony, beryllium, cadmium, chromium, cobalt, manganese, mercury, tantalum, molybdenum, niobium, titanium, tungsten, vanadium and zicronium - are vital to modern industry although they are only produced in relatively small quantities.

3.3.6 Sintered metals

Sintered metals are those produced by the techniques of powder metallurgy and they posses very different properties and structures from those of metals that have been cost and worked using conventional techniques.

3.4 PHASE AND STRUCTURES IN STEELS

Pure iron is a relatively soft material and is hardly of any commercial use in that state.

Ferrous alloys are of the utmost importance in engineering, and are classified on the basis of the percentage of carbon and other alloying elements present.

The general term steel has been used to describe iron-carbon alloys with a carbon content between 0.1% and 1.7% (some of the literature-maximum of about 2%). These alloys may contain other element and can be produced in various grades possessing a wide range of properties. In order to understand the structural change which occur and the phases that are present when iron-carbon alloys are cooled, it is necessary to interpret the various regions of the phase diagram (iron-carbon equilibrium diagram) as shown in Fig. 3.1. Certain special names are used to describe the constituents that may be present in different steels. These terms will now be defined and will be used throughout the remainder of the chapter.

3.4.1 Ferrite

Alpha (α) iron is commonly called ferrite. It is the softest of all materials in the diagram. Pure α-iron is a soft, ductile metal with little mechanical strength, and is a ferromagnetic material at temperatures below 768°C (the Curie temperature). The loss of ferromagnetism above 768°C is a reversible process on cooling. Ferrite is name given to the bcc solid solution of iron and carbon exists at temperatures below 908°C. The maximum carbon that exists at temperature below 908°C.

3.4.2 Austenite

Austenite is the name given to fcc solid solution. The allotropic change of pure iron from α-form to the γ-form at 908°C causes reduction in volume, because the fcc structure has a denser atomic packing. γ-iron is an alternative term, but this is reserved for the pure fcc iron which is the stable form between 908°C and 1388°C. The austenitic solid solution can accept up to 1.7% carbon (some of the literature maximum of about 2%), it is relatively soft and ductile material which is well suited to most fabrication processes.

3.4.3 Cementite

Cementite is the name give to iron carbide (Fe_3C), which has a crystal lattice containing iron and carbon atoms in a three to one ratio. Compared to ferrite or austenite, cementile is hard and brittle constituent. It is relatively weak as it cannot adjust to stress concentrations, but when combined with ferrite produces a strong steel.

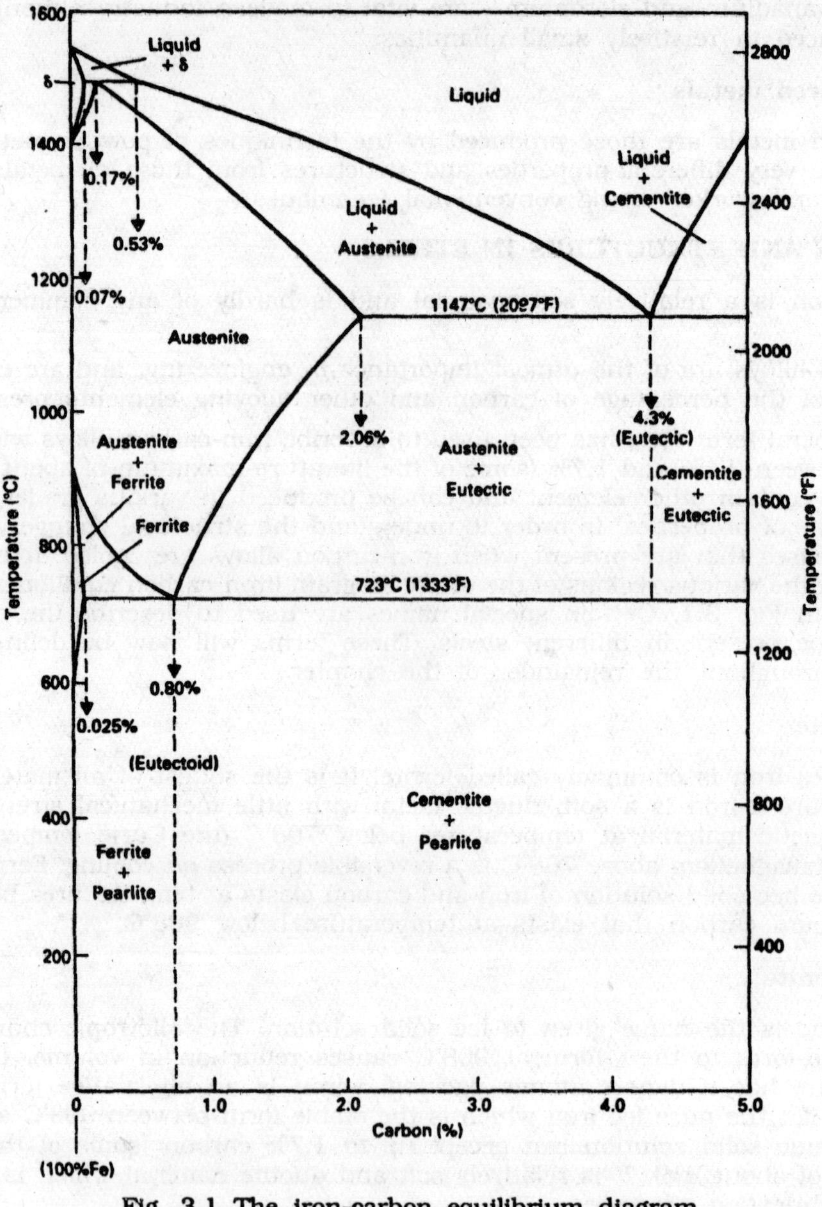

Fig. 3.1 The iron-carbon equilibrium diagram.

3.4.4 Eutectic (Ledeburite)

It contains 4.3% carbon but is rarely seen in slowly cooled alloy since it breaks down, due to its unstable nature, to other phases. It consists of a mixture of austenite and cementile known as ledeburite. It solidifies isothermally at a temperature of 1147°C [Fig. 3.2].

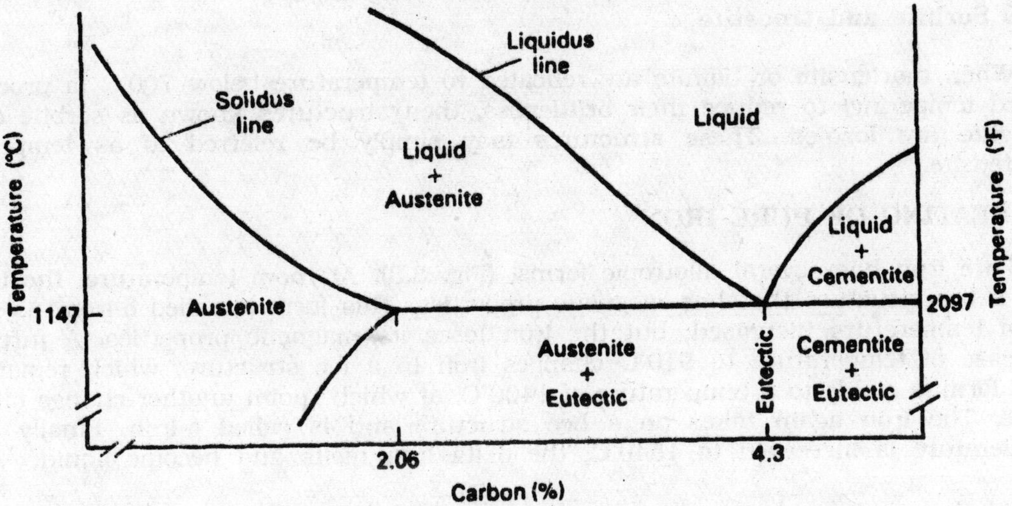

Fig. 3.2 The Eutectic region

3.4.5 Pearlite

This is the eutectoid formed within the alloy system by the simultaneous precipitation of ferrite and cementite from austenite at the lower critical temperature of 723°C. When viewed under a microscope, the mixture appears in lamellar layers of cementite within a field of white ferrite. The name pearlite has its origin in the fact that microstructure resembles that of mother-of-pearl. Pearlite is a harder and much less ductile.

3.4.6 Martensite

If austenite is rapidly cooled it does not have time to change to ferrite and cementite; then all the carbon remains in the solid solution. The mixture is ferrite highly supersaturated with dissolved carbon, and it is known as martensite. On cooling, the fcc structure of austenite changes to a bcc, but in order to accommodate all the carbon this is a body centered tetragonal (bct) structure for martensite, rather than the bcc structure of ferrite solutions. Martensite is hard, strong and brittle.

3.4.7 δ-iron

δ-Iron is the name given to the bcc pure allotrope of iron which is stable between 1388°C and its melting point of 1535°C. The term δ-iron is also used to describe the solid solutions of iron-carbon at these higher temperatures, with a maximum solubility of 0.09% carbon. It is sometimes referred to as δ-ferrite.

3.4.8 Bainite

If austenite is cooled at a slower rate than that required to produce martensite, but at a rate too fast to allow pearlite to form, then bainite is the decomposition product. Bainite is also formed by isothermal transformation.

3.4.9 Sorbite and troostite

When martensite or bainite are reheated to temperature below 700°C (a process called tempering) to reduce their brittleness, then structures known as sorbite and troostite are formed. These structures may simply be referred to as tempered martensite.

3.5 HEATING OF PURE IRON

Pure iron has several allotropic forms, (Fig. 3.3). At room temperature, the iron has a bcc structure that has magnetic properties. This form is called δ-iron. As the room temperature increased, but the iron loses its magnetic properties. A further increase of temperature to 910°C changes iron to a fcc structure, which remains. This form is stable to a temperature of 1400°C, at which point another change takes place. The iron again takes on a bcc structure and is called δ-iron, Finally the temperature is increased to 1540°C, the delta iron melts and become liquid.

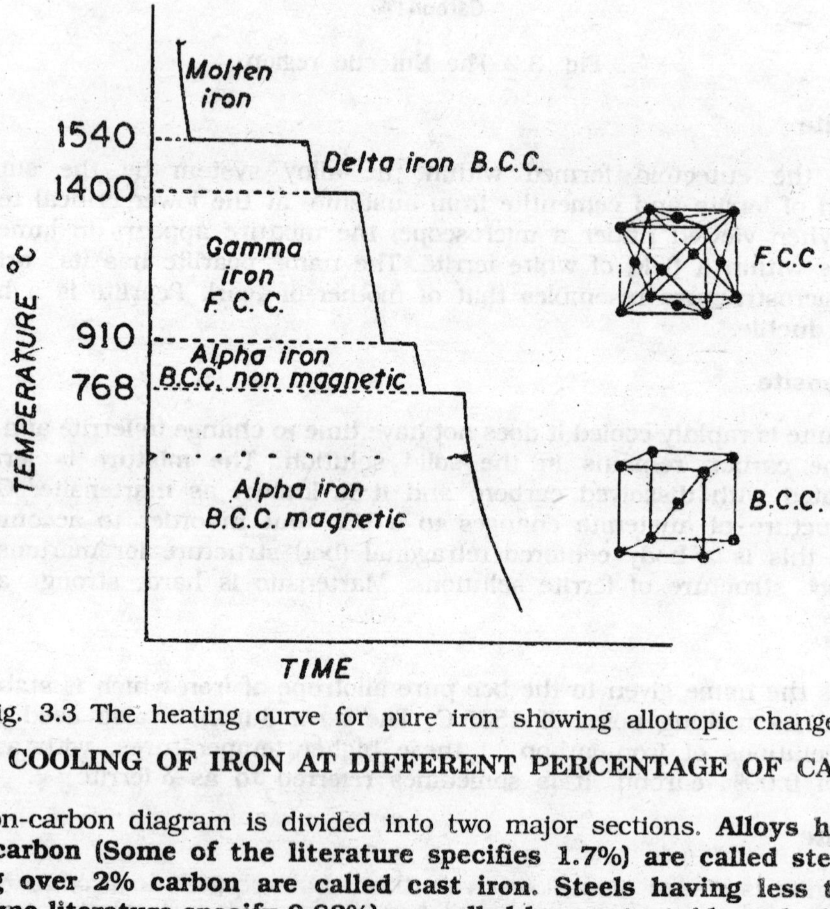

Fig. 3.3 The heating curve for pure iron showing allotropic changes

3.6 SLOW COOLING OF IRON AT DIFFERENT PERCENTAGE OF CARBON

The iron-carbon diagram is divided into two major sections. **Alloys having less than 2% carbon (Some of the literature specifies 1.7%) are called steels; alloys containing over 2% carbon are called cast iron. Steels having less than 0.8% carbon (some literature specify 0.83%) are called hypoeutectoid steel; those above 0.8% carbon are called hypereutectoid steels.**

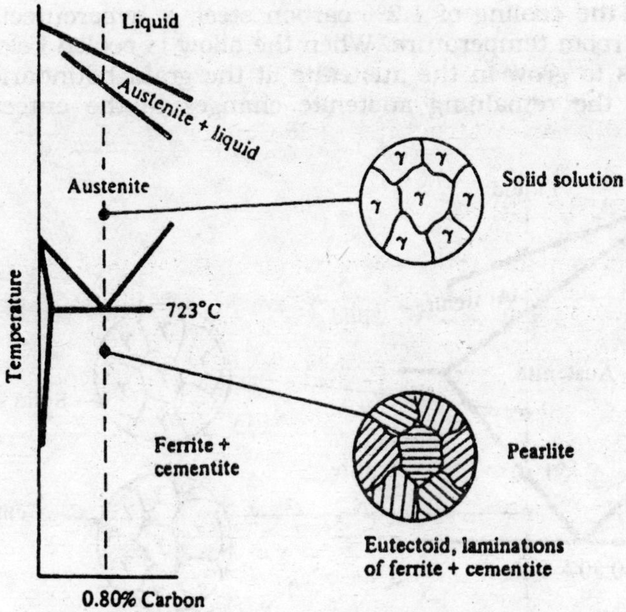

Fig. 3.4. Slow cooling of 0.8% carbon steel.

Fig. 3.4 shows the cooling from the liquid of an alloy with 0.8% carbon. Fig. 3.5 shows the cooling of 0.4% carbon steel, a hypoeutectoid steel, from the austentite phase to room temperature. When the alloy is cooled below temperature T_1, crystals of ferrite start to grow in the austenite. The ferrite tends to grow at the grain boundaries of the austenite crystals. **At 723°C the remaining austenite changes to the eutectoid structure, i.e. pearlite.**

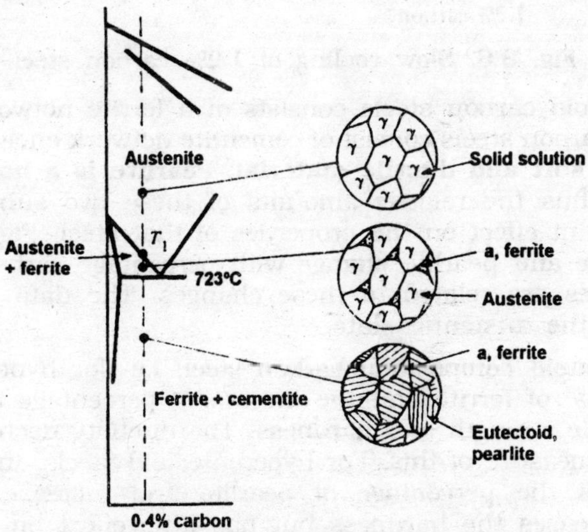

Fig. 3.5. Slow colling of 0.4 carbon steel.

Fig. 3.6 Shows the cooling of 1.2% carbon steel, a hypereutectoid steel, from the austenite phase to room temperature. When the allow is cooled below the temperature T_1, cementite starts to grow in the austenite at the grain boundaries of the austenite crystals. At 723°C the remaining austenite changes to the eutectoid structure, i.e. pearlite.

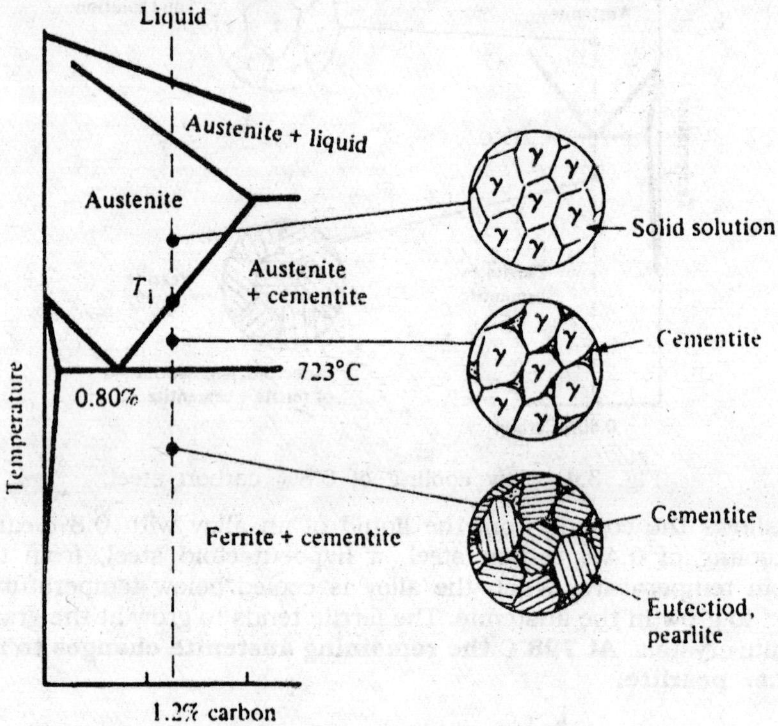

Fig. 3.6. Slow cooling of 1.2% carbon steel.

Thus hypoeutectoid carbon steels consists of a ferrite network enclosing pearlite and hypoeutectoid carbon steels consist of cementite network enclosing pearlite. **Ferrite is a comparatively soft and ductile material. Pearlite is a harder and much less ductile material.** Thus the relative amounts of these two substances in a carbon steel have a significant effect on the properties of that steel. Fig. 3.7 shows how the percentages of ferrite and pearlite change with percentage carbon and also how the mechanical properties are related to these changes. The data refers only to steels cooled slowly from the austentic state.

Up to the euctectoid composition carbon steel, i.e. for hypoeutectoid steels, the decreasing percentage of ferrite and the increasing percentage of pearlite results in an increase in tensile strength and hardness. The ductility decreases, the elongation at fracture being a measure of this. For hypereutectoid steels, increasing the amount of carbon decreases the percentage of pearlite and increases the percentage of cementite. This increases the hardness but has little effect on the tensile strength. The ductility also changes little.

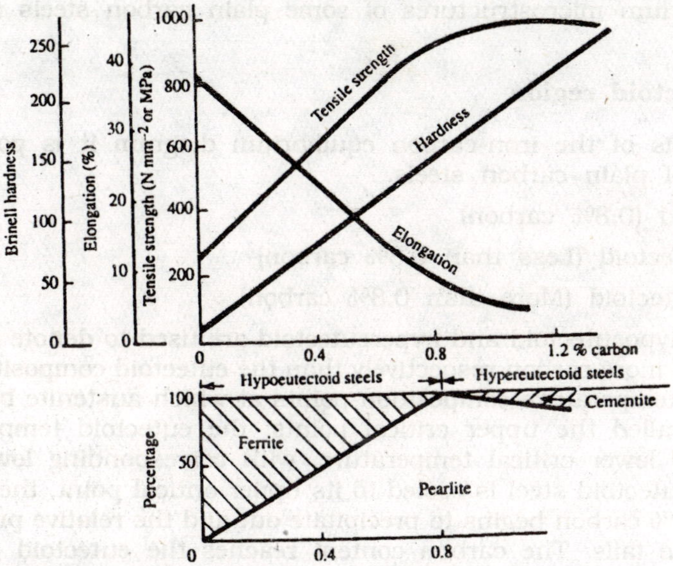

Fig. 3.7. The effect of carbon content on the structrue and properties of steels.

3.7 PLAIN CARBON STEELS

A plain carbon steel an alloy containing carbon as its principal alloying element, other alloying elements such as manganese, sulphur, phosphorous, silicon and nickel being present in very small amounts only.

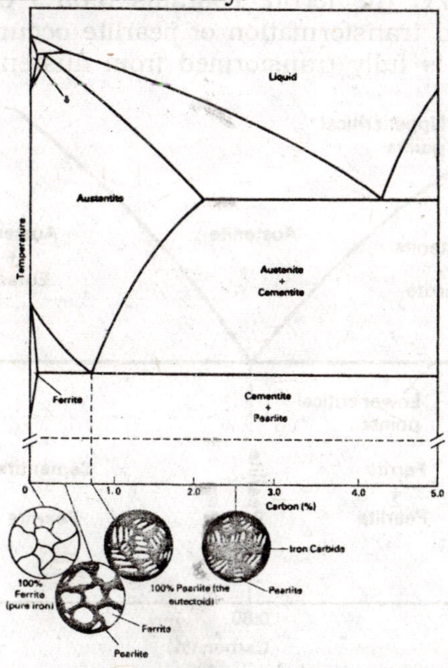

Fig. 3.8 Equilibrium microstructures of plain Carbon Steels at room temperature.

The equilibrium microstructures of some plain carbon steels are shown in Fig. 3.8.

3.7.1 The eutectoid region

On the basis of the iron-carbon equilibrium diagram it is possible to describe three classes of plain carbon steels.

(1) Eutectoid (0.8% carbon)

(2) Hypoeutectoid (Less than 0.8% carbon)

(3) Hypereutectoid (More than 0.8% carbon)

The terms hypoeutectoid and hypereutectoid are used to denote steels that contain less carbon and more carbon respectively than the eutectoid composition (0.8% carbon). In Fig. 3.9 the temperature composition points at which austenite begins to transform to ferrite are called the upper critical points; the eutectoid temperature (723°C) is known as the lower critical temperature, with corresponding lower critical points. When a hypereutectoid steel is cooled to its upper critical point, then cementite (Fe_3C) containing 6.67% carbon begins to precipitate out and the relative proportion of carbon in the austenite fails. The carbon content reaches the eutectoid composition (0.8% carbon) at a temperature of 723°C; the transformation of austenite to cementite and pearlite is then complete. For hypoeutectoid steel containing between 0.03% and 0.8% carbon, upon cooling to the upper critical point ferrite begins to precipitate out. The ferrite can only contain a small amount of carbon in solid solution, and therefore the relative proportion of carbon remaining in the austenite is increased. As the temperature falls the carbon dissolved in the ferrite increases, as does the carbon in the austenite. At 723°C the ferrite contains 0.03% carbon, the austenite is of eutectoid composition and transformation of pearlite occurs. An alloy which contains less than 0.03% carbon is fully transformed from austenite to ferrite (all carbon in

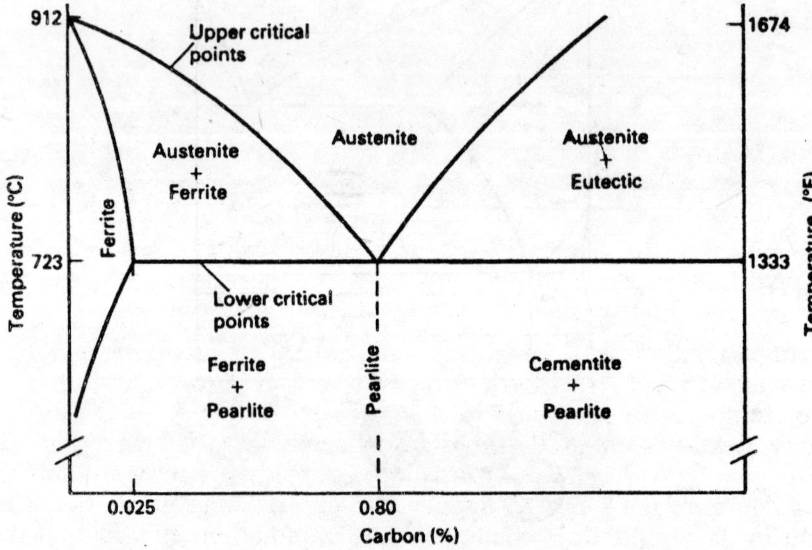

Fig. 3.9 The eutectoid region

solid solution) before 723°C is reached. At room temperature the maximum solubility of carbon in ferrite is 0.006%. Therefore, if an alloy contains between 0.006% and 0.03% carbon, a further structural change occurs when it is cooled to some temperature below 723°C. This situation occurs when the lower temperature solubility of carbon is reached; the carbon then precipitates out of the ferrite as pearlite.

3.7.2 Time-Temperature-Transformation diagrams (T T T diagrams)

The rapid cooling of steel does not create equilibrium conditions, therefore, the metastable phases may be formed. However, the rate of cooling is an important consideration when producing different steels, and an alternative diagram is used to interpret the effects which occur. These diagrams are known as Time- Temperature-Transformation diagram (T T T diagrams), isothermal transformation curves, 'S' curves, or 'C' curves (due to their general shape). The curves are different for each steel, although their general shape is similar. An example is shown in Fig. 3.10 for a plain carbon steel.

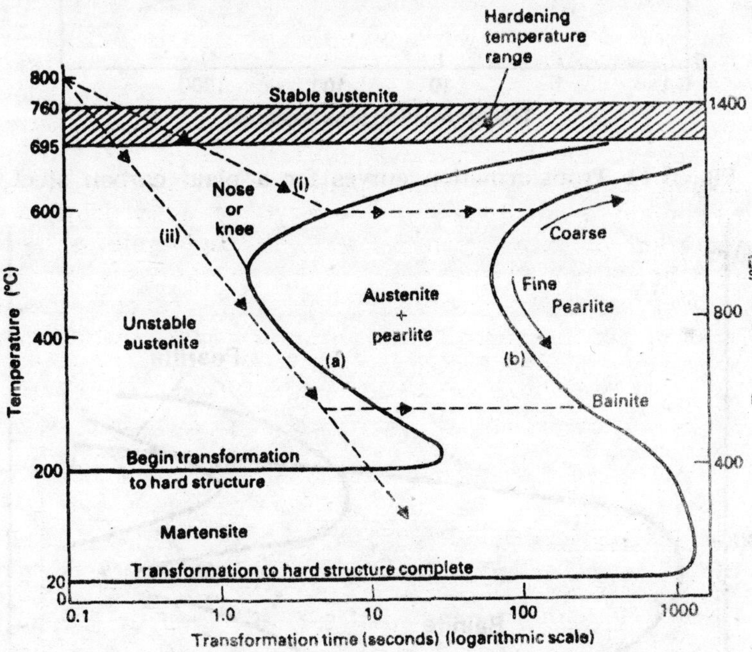

Fig. 3.10 Typical Time-Temperature-Transformation diagram for a plain carbon steel. (a) Transformation begins (b) Transformation ends.

Considering Fig. 3.10 at temperature above 723°C, austenite is the stable form of this plain carbon steel. This temperature depends upon the actual carbon content. The range of temperatures above 723°C to which the steel can be heated and held as preliminary heat treatment. If the steel is cooled very slowly, then the austenite transforms to pearlite very slowly and a final coarse structure is obtained with more rapid cooling, as shown by dotted line (i) in Fig. 3.10, the grain structure is refined and fine pearlite is obtained. If cooling is very rapid, then the cooling rate is shown by dotted line (ii) in Fig. 3.10, the critical cooling rate is such that the 'knee' of

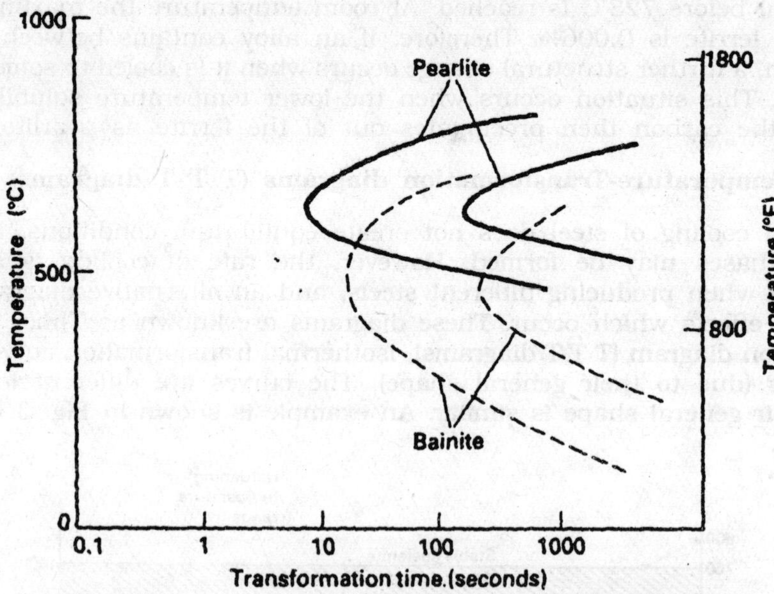

Fig. 3.11 Transformation curves for a plain carbon steel.

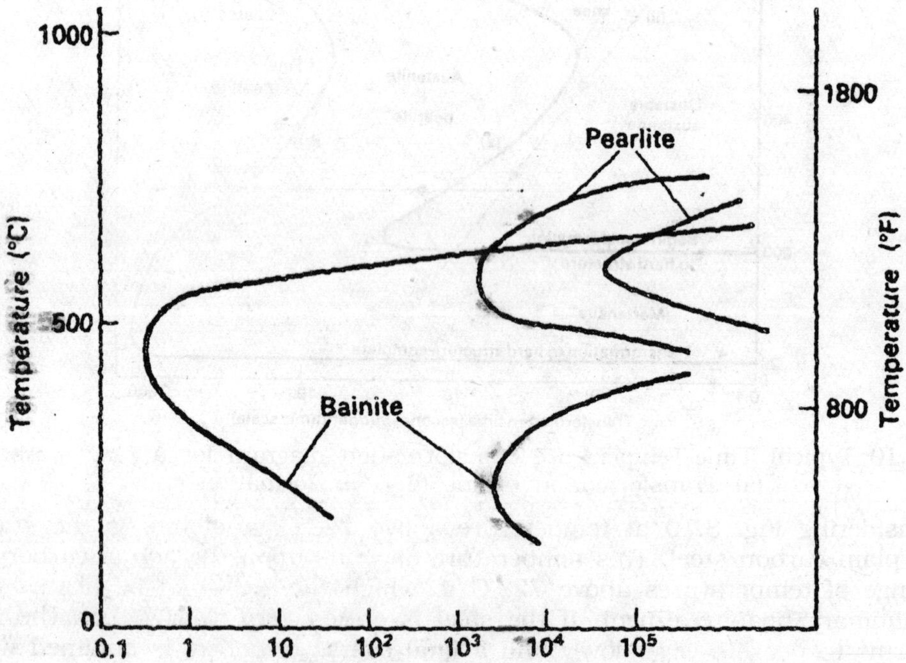

Fig. 3.12 Transformation curves for an alloy steel.

the curve is avoided and martensite rather than pearlite is formed. If this rapid cooling curve is followed (curve ii) but is arrested before martensite is formed (and below the knee of the curve), then if the temperature is held constant the product is bainite rather than pearlite. This process is known as austempering. Bainite is an intimate mixture of carbide and ferrite (not laminations), it is formed because the transformation curves (a and b) of Fig. 3.10 are really a combination of two sets of curves. This situation is shown in Fig. 3.11 and 3.12 for a plain carbon steel and an alloy steel. Pearlite formation requires sufficient time for diffusion of the carbon and iron atoms, whereas bainite formation requires sufficient time for the shearing of the fcc structure to a bcc lattice as well as carbon precipitation.

TTT diagrams can be used for the selection and specification of several types of heat treatment.

3.7.3 Classification of plain carbon steel and their applications and uses.

The classification of plain carbon steels and their applications and uses are summarized in Table 3.1.

Table 3.1. Some applications and uses of plain carbon steels.

Type of Steel	% Carbon	Application and uses
Dead Mild	0.05-0.15	Chains, Stamping, rivets, nails, seam welded pipes, tin plate, automobile body steel, and material subject to drawing and pressing.
Mild Steels	0.10-0.20	Structural steels, screws, drop forgings, case hardening steel
	0.20-0.30	Machine and structural work, gears, free cutting steels, shafting and forgings.
Medium carbon	0.30-0.40	Connecting rods, shafting, axles, crane hooks, forgings.
	0.40-0.50	Crankshafts, axles, gears, shafts, dieblocks, rotors, tyres, skip wheels
	0.50-0.60	Loco tyres, rails, wire ropes
High Carbon	0.60-0.70	Drop hammer dies, saws, screwdrivers
	0.70-0.80	Band saws, anvil faces, hammers, wernches, laminated springs, cable wire, large dies for cold presses.
	0.80-0.90	Cold chisels, shear blades, punches, rock drills
Tool steels	0.90-1.10	Axes, knives, drills, taps, screwing dies, picks
	1.10-1.40	Ball bearings, files, broaches, razors, boring and finishing tools, machine parts where resistance to wear is essential.

Plain Carbon Steel : Plain carbon steel contains a maximum of 1.5% carbon, 1.5% manganese and 0.5% silicon. and only traces of other elements. The carbon content affects the hardness and strength of the steel. Plain carbon steels often classified into the following three main categories.

(1) Mild steel or low carbon steel : This type of steel contains a maximum of between 0.25 and 0.3% carbon. It is easily machined and welded, but does not respond to heat treatment. It is used for constructional applications and can be forged or rolled to form angle, channel, plate, bar or rod.

(2) **Medium Carbon Steel :** Steel with a carbon content between 0.25% and 0.65% is termed a medium carbon steel. The hardness is often improved by heat treatment. It is shock resisting steel used for crankshafts and axis.

(3) **High Carbon Steel :** This is a plain carbon steel containing 0.6% and 1.5% carbon. It is always used in a hardened (or tempered) condition, but it is also brittle. These steels are also known as tool steels because of their applications. A silver steel has a high carbon content; it contains no silver but has a bright appearance. Silver steel is produced as bar with very close limits of accuracy on the diameter.

This classification of plain carbon steels is of more direct use to the engineer. Table 13.1 gives uses of plain carbon steels.

3.7.4 Limitations of plain carbon steels

Plain carbon steels limits on the engineering applications because :

(1) A high-strength steel can only be obtained by increasing the carbon content to such a level that the material becomes brittle. **High strength cannot be obtained with good ductility and toughness.**

(2) **Hardness requires water quenching.** The severity of this rapid rate of cooling often leads to distortion and cracking of the steel.

(3) Large sections cannot be hardened uniformly. **The hardness depends on the rate of cooling and this will vary across a large section.**

(4) **Plain carbon steels have poor resistance to corrosion and oxidation at high temperatures.**

3.8 ALLOY STEELS

The term alloy steel is used to describe those steels to which one or more alloying elements, in addition to carbon, have been deliberately added in order to modify the properties of the steel. If alloying elements other than carbon present only in small amounts (e.g. manganese upto 0.8%, silicon upto 0.3%, etc.) then the steel is usually called **a low alloy steel, or plain carbon steel.** These are added during the steel making in order to overcome the effects of the two main impurities - sulphur and oxygen that are invariably present.

3.8.1. Low alloy steel

Low alloy steel have a microstructure similar to plain carbon steels, and require the same heat treatment. They have comparable applications and similar carbon contents. Low alloying steels usually contain upto 3 or 4% of one or more alloying elements to improve the strength, toughness and hardenability. Nickel improves the resistance to fatigue.

The **term low alloy is used** for alloy steels when the alloying additions are **less than 2%, medium alloy** when the additions are between **2 and 10%**, and **high alloy** when they are **above 10%**. In all cases the **amount of carbon present is less than 1%**. Common elements that are added are **aluminium, chromium, cobalt, copper, lead, manganese, molybdenum, nickel, phosphorus, silicon, sulphur, titanium, tungsten and vanadium.**

There are a number of ways in which the alloying elements can have an effect on the properties of the steel. **The main effects are to:**

(1) Solution harden the steel : Most of the elements used in alloy steels form substitutional solid solutions with the iron. This increases tensile and impact strengths.

(2) Form carbides : They may form stable, hard carbides which may, if in a appropriate form such as fine particles, increase the strength and hardness. Manganese, chromium and tungsten have this effect.

(3) Form graphite : They could cause the breakdown of cementite and lead to the presence of graphite in the alloy. Silicon and nickel have this effect. The result is a decrease in strength and hardness.

(4) Stablize austenite and ferrite : They may lower the temperature at which austenite is formed on heating the steel. Manganese, nickel, copper and cobalt have this effect. With manganese about 11 to 14% produces what is known as an **austentic steel**. Such steels have relatively good hardness combined with ductility and so are tough. They may increase the temperature at which austenite is formed on heating the steel. Chromium, molybdenum, tungsten, vanadium, silicon, and aluminium have this effect. This raises the temperature to which the steel has to be heated for hardening. If a sufficiently high percentage of one of these element is added to the steel, the transformation from ferrite to austenite may not take place before the steel reaches its melting point temperature. Such steels are known as ferrite steel, 12 to 25% of chromium with a 0.1% carbon steel gives a such a type of steel. Such a steel cannot be hardened by quenching.

(5) Change the critical cooling rate : The critical cooling rate is the minimum rate of cooling than has to be used if all the austenite is to be changed into martensite. With the critical cooling rate, or a higher rate of cooling, the steel has maximum hardness. With slower rates of cooling a less hard structure is produced. **Most alloying elements reduce the critical cooling rate**. The effect of this is to make air or oil quenching possible, rather than water quenching. It also increases the hardenability of steels.

(6) Improve corrosion resistance : They can improve the corrosion resistance. Some elements promote the production of adherent oxide layers on the surfaces of the steel and so improve its corrosion resistance. **Chromium is particularly useful in this respect. If it is present in a steel in excess of 12% the steel is known as stainless steel because of its corrosion resistance.** Copper is also used to improve corrosion resistance.

(7) Change grain growth : They can influence grain growth. Some elements accelerate grain growth while others decrease grain growth. **The faster grain growth leads to large grain structures and consequently to a degree of brittleness. The slower grain growth leads to smaller grain size and so to an improvement in properties. Chromium accelerates grain growth** and thus care is needed to heat treatment of chromium steels to avoid excessive grain growth. **Nickel and vanadium decrease grain growth.**

(8) Improve machinability : They can effect the machinability of the steel. Sulphur and lead are elements that are used to improve the chip formation properties of steels.

The general effects and functions of particular alloying elements and the types of steels produced are summarized in Table 3.2.

Table 3.2. Effects and functions of alloying elements in steels.

S.N.	Elements	Main effects	Main functions
1	Aluminium	Ferrite stablizer	Aids nitriding
2	Chromium	Increase strength, fatigue resistance and hardness; forms hard and stable carbides; raises the critical temperature range; increases hardenability; amounts in excess of 12% Cr render the steel stainless After heat treatment it produces a uniform structure and a uniform hardness at greater depths, also improved tensile and elastic properties. Also used for chromium plating steels.	1.0% to 1.5% Cr in medium and high-carbon steels for gears, axies, shafts and springs, ball bearings and metal-working rolls; 12% to 30% Cr in mrtensitic and feritic stainless steels; also used in conjunction with nickel (see below)
3	Cobalt	Austenite stablizer	Improves strength at high temperatures
4	Copper	Austenite stablizer	Improves corrosion resistance
5	Lead	Improves free-cutting properties, especially for automatic machining	Used in carbon steels.
6	Maganese	Increases the strength and hardness and forms a carbide; increases hardenability; lowers the critical temperature range; in sufficient quantity produces an austenitic steel; always present in a steel because it is used as a deoxidizer. Induces sulfur impurity into the slag. Neutralizes effects of impurities in welds. Produces a uniform hardness of greater depth after heat treatment. Makes steel easier to forge due to improved ductility and flow characteristics. Can also produe free-cutting properties.	Pearlitic steels (up to 2% Mn) with high hardenability used for shafts, gears, and connecting rods. 13% Mn in Hadfield's steel – a tough austenitic steel.
7	Molybdenum	Strong carbide forming element; improves high-temperature; hardened creep resistance; reduces temper-brittleness in Ni-Cr steels.	Not normally used alone; a constituent of high-speed tool steels and creep-resistant steels; up to 0.5% Mo often added to pearlitic temper-brittleness.
8	Nickel	Marked strengthening effect; lowers the critical temperature range; increases hardenability; improves resistance to fatigue; strong graphite-forming tendency; stabilizes austenite when present in sufficient quanityt. Improved ductility and abrasion resistance. Lowers the critical points at which structural changes occur during heat treatment; therefore, less heat required and	0.3% to 0.4 C with up to 5% Ni used for crankshafts and axies, and other parts subject to fatigue.

		more gradual cooling, also les distortion and cracking. Used as corrosion-resistant plating on steels. Austenitic nickel steels quickly work harden.	
9	Nickel and chromium	Frequently used together in the ratio Ni:Cr = 3:1 in pearlitic steels; good effects of each element are additive; each element counteracts disadvantages of the other; also used together for austenitic stainless steels.	0.15% C with Ni and Cr used for case carburizing; 0.3% C with Ni and Cr used for gears, shafts, axles and connecting rods, 18% (or more) or chromium and 8% (or more) of nickel produces austenitic stainless steels.
10	Phosphorus	Solid solution hardening Improves chip formation	Improves machinability
11	Silicon	Strengthens ferrite; raises the critical temperature range; a strong graphitizing tendency; always present because it is used (with manganese) as a deoxidizer.	Silicon steel (0.07%, C; 4% Si) used for transformer cores; used with chromium (3.5% Si; 8% Cr) for its high-temperature oxidation resistance in internal combustion engine valves.
12	Sulphur	Improves chip formation	Improves machinability
13	Titanium	Strong carbide-forming element	Not used alone, but added to high-speed steels, and to some pearlitic chromium steels.
14	Tungsten	Forms hard and stable carbides; raises the critical temperature range and tempering temperatures; hardened tungsten steels resist tempering up to 600°C (1112°F)	Major constituent in high-speed tool steels; also used in some permanent magnet steels.
15	Vanadium	Strong-carbide-forming element; scavenging action and produces clean, inclusion-free steels.	Not used alsone, but it is added to high-speed steels, and to some pearlitic chromium steels

3.8.2 High alloy steels

The structures of high alloy steels and the heat treatment processes that are required are very different from plain carbon steels. These are special steels that have particular characteristics such as corrosion resistance and low temperature applications. The uses of certain high alloy steels are as follows :

(1) High speed tool steels

The addition of tungsten and chromium to a high carbon steel causes the formation of very hard and stable carbides, and a subsequent increase in the critical temperature and the softening temperature. These steels can then be used to produce hard wearing, metal cutting tools which can be operated up to 600°C. A typical composition is 18% tungsten, 4% chromium, 1% vanadium and 0.8% carbon, which is known as 18/4/1 steel.

(2) Wear resisting steels

An austenitic manganese steel contains 12% to 14% manganese, 0.75% silicon and 1% carbon. It may also contain some chromium or vanadium which from carbides and improve the strength. The high manganese content makes the steel austenitic and magnetic at all temperatures. It can be obtained as castings, forgings or rolled sections, but will work hardened if machined. It has excellent resistance to abrasion and is used in crushing machinery and pneumatic drill bits. An alternative hard-wearing steel contains 1.4% chromium, 0.45% manganese and 1% carbon.

(3) Corrosion resistant steels (Stainless Steels)

The development of stainless steels has been mainly due to their corrosion-resisting properties. However, many of these steels can also be used at high temperatures where alloying additions to the steel can reduce the main problems of loss of strength, creep and oxidation or chemical attack. The classification is same for both stainless steels and heat resisting steels.

Table 3.3. Structure and applications of stainless steels

S.N.	Type	Structure	Compositions	Special Applications
(1)	Ferritic	Ferritic grains plus	12% to 25% chromium and less than 0.1% carbon	Where outstanding corrosion resistance and excellent formability are required, e.g. for many deep drawing applications
(2)	Martensitic	Martensitic Structures if oil quenched from above upper critical Temperature	(a) Stainless irons 12% to 13% chromium and 0.07% to 0.1% carbon	(a) Turbine blades, rivets, split pins, cutlery
			(b) Stainless steels 13% chromium and 0.25% to 0.3% carbon	(b) Surgical instruments springs, ball bearings.
			(c) High-chromium stainless steels 16 to 18% chromium 2% nickel and 0.1% carbon.	(c) Pump shafts, regulator valves, aircraft fitting, turbine blades.
(3)	Austenitic	Fully austenitic if quenched from above upper critical Temperature (UCT)	Variations of the 18% chromium, 8% nickel basic alloy usually about 0.1% carbon	Chemical plant construction; decorative purposes; domestic applications such as cutlery, saucepans, etc.

(a) Ferritic : Low carbon content (maximum 0.1%) and between 12% and 25% chromium. The small amount of carbon remains dissolved in the iron up to the melting point, and austenite cannot be formed. Therefore, this steel cannot be hardened by heat treatment but may be strengthened by work hardening.

(b) Martenistic : Higher carbon content (between 0.1% and 1.5%) restoring the ferrite-to-austenite transformation and can be hardening by heat treatment. It contains in between 12% and 18% chromium.

(c) Austenitic : It contains both chromium and nickel. The carbon content is kept below 0.15% to limit carbide formation which would reduce the corrosion resistance. Cannot be heat treated because the austenitic structure is retained due to the alloying elements. Carbides can form if the steels are slowly cooled from high temperatures, or reheated to 500°C to 700°C. These carbide particles can cause weld decay; small amounts of stabilizers such as titanium or niobium should be present.

Table 3.3 shows the approximate compositions, structures and typical applications of these three groups of stainless steels.

The effects of the main alloying elements on the heat resisting properties of steels, and their classifications, are summarized in Table 3.4.

The principal effects certain alloying elements are summarized in Table 3.5.

Table 3.4 Types of heat resisting steels

Property required at high temperature	Alloying element
Strength	Nickel or chromium, producing an austenitic structure, Nickel also retards grain growth
Creep resistance	Small amounts of titanium, aluminum and molybednum, with carbon.
Resistance to oxidation and chemical attack	Silicon or chromium

Type	Composition (all %)					Maximum temperature	Comments
	C	Cr	Ni	Si	Mn	(°C/°F)	
Ferritic	0.06	21	—	0.4	0.8	900/1652	Work-hardened steel, resistant to sulfur gases.
	0.1	29	—	1.2	1.3	1100/2012	Hardened by heat treatment, easily machined, welding not recommended
Martenstic	0.4	11.5	—	0.3	0.5	750/1382	
Austentic	0.12	18.5	8.5	1.0	0.7	350/662	Easily welded, difficult to machine
	0.2	23	11.5	1.6	0.4	1105/2021	3%
	0.15	13.5	63.0	0.8	1.3	1100/2012	Very low coefficient of thermal expansion

Table 3.5. Effects of alloying elements.

Elements	Typical ranges in alloys steels (%)	Principal effects
Aluminum	< 2.0	Aids nitriding; restricts grain growth; removes oxygen in steel melting
Sulfur	< 0.5	Aids machinability; reduces weldability and ductility
Chromium	0.3 to 4.0	Increases resistance to corrosion and oxidation; increases hardenability; increases high temperature strength; can combine with carbon to form hard wear- resistant microconstituents.

Nickel	0.3 to 5.0	Promotes an austenitic structure; increases hardenability; increases toughness.
Copper	0.2 to 0.5	Promotes a tenacious oxide film to aid atmospheric corrosion resistance
Manganese	0.3 to 2.0	Increases hardenability; promotes an austenitic structure; combines with sulfur and reduces it adverse effects.
Silicon	0.2 to 2.5	Remove oxygen in steel making; improves toughness increases hardenability
Molybdenum	0.1 to 0.5	Promotes grain refinement; increases hardenability; improves high-temperature strength
Vanadium	0.1 to 0.3	Promotes grain refinement; increases hardenability; will combine with carbon to form wear-resistant microconstituents.

3.9 CAST IRONS

The term cast iron arises from the method by which the iron is produced. Pig iron is remelted in a furnace and the properties of the iron modified by the addition of other materials. The resulting iron is them cast, hence the term cast iron. Cast irons have between about 2% and 4% carbon and often significant amounts of silicon, as well as smaller amounts of other elements. **Cast irons are used widely because of their ease of melting and hence the production of components by casting.** A cast iron is classified as either hyper-eutectic or hypo- eutectic. However, this classification has little industrial significance.

3.9.1 Types of Cast iron

Examination of the iron-carbon equilibrium diagram (Fig. 13.1) reveals that the slow cooling of cast irons should result in structures that consist essentially of cementite and pearlite. However, in practice the actual rate of cooling and the presence of other alloying elements such as silicon, manganese phosphorous and sulphur play a determining role in the type of microstructure, formed in any particular grade of cast iron. The two extremes of microstructure found in cast irons, which are rarely achieved in practice, are (1) all of the carbon is combined as cementite, the resulting microstructure consisting entirely of cementite and pearlite, and (2) all of the carbon exists as free graphite in a matrix or ferrite. The following four basic types of cast irons are usually produced.

(1) Grey Cast Iron : The amount of grey cast iron produced for exceeds that of any other type of cast iron. The microstructure of a typical grey cast iron reveals a large amount of free carbon in the form of both flake and rosette graphite in a matrix of, usually, pearlite and ferrite. **When broken, this material exhibits a grey colour on the fracture surface, the brittleness exhibited being due to the weakening effects of the free graphite. Grey Cast iron is typically weak in tension, fairly soft, brittle, strong in compression, and possesses excellent properties.**

(2) White Cast Iron : White cast iron, so named because a fractured surface is white in colour, is a result of fast cooling and has a structure composed entirely of cementite and pearlite. Thus, all of the carbon is in the combined form (cementite) and the iron is extremely hard and virtually unmachinable. White cast iron is always low in silicon since silicon is the most powerful graphitising agent that can be added to a cast iron melt.

White cast iron is seldom produced except as an intermediate step in the production of malleable irons. However, thin layers of white iron are often produced on the surface of grey iron castings by using metal "chill bars" in the moulding sand close to the metal surface; this produces a casting with a strong and hard surface and a ductile, shock resistant core.

(3) Malleable cast irons : Malleable cast irons are produced by the heat treatment of white cast irons. Three forms of malleable iron occur :

(i) blackheart,

(ii) white heart, and

(iii) pearlitic

Malleable irons have better ductility than grey cast irons and this combined with their high tensile strength makes them useful materials.

In the blackheart process white iron castings are heated in a non-oxidizing atmosphere to 900°C and soaked at that temperature for two days or more. This causes the cementite to break down. The result is spherical aggregates of graphite in austenite. The casting is then cooled very slowly, resulting in the austenite changing into ferrite and more graphite (Fig. 3.13).

Fig. 3.13. Production of blackheart malleable iron.

The whiteheart process also involves heating white iron casting to about 900°C and soaking at that temperature. But in this process the castings are packed in consisters with haematite iron ore. This gives an oxidizing atmosphere. Where the casting is thin, the carbon is oxidized forming a gas and so leaves the casting. In the thicker sections of the casting only the carbon in the surface layers leaves. The result, after very slow cooling, is a ferritic structure in the thin sections of the casting and, in the thick sections, a ferrite outer layers with a ferrite plus pearlite inner core.

Pearlite malleable iron is produced by heating a white iron casting in a non-oxidizing atmosphere to 900°C and then soaking at that temperature. This causes the cementite to break down and given spherical aggregates of graphite in austenite, as with blackheart iron. However, a more rapid cooling is used a pearlite structure is produced. This pearlite malleable iron has a higher tensile strength than blackheart iron.

Pearlite malleable irons can be produced by other methods. One method involves adding 1% manganese to the iron. The manganese inhibits the production of graphite. **The higher strength pearlitic malleable irons are produced by quenching the iron from 900°C and then tempering.**

(4) Nodular Cast Irons : Nodular or spherodial graphite (S.G.) iron is produced by inoculating white cast iron melts with small amounts of such materials as nickel-magnesium (2%), ferro-silicon, or calcium silicide just prior to pouring. The resulting structures may vary from almost completely pearlitic matrixes in which spheroids or nodules of carbon are imbedded. Up to 15% elongation can be present in such irons, which are used where shock-resistance is required. Pearlitic S.G. irons can be surface hardened by a flame hardening technique, and complete hardening can be achieved by quenching; in the later case the iron can also be tempered to remove brittleness.

(5) Austempered ductile irons : Ductile irons are generally used in the as-cast condition or possible after an annealing or normalizing treatment. The properties can, however, be considerably improved by an austempering heat treatment. This treatment involves heating the casting to 850 to 950°C, the immersing it in a bath of liquid held at a temperature which is usually in the range 235 to 425°C. It is held at this temperature for up to four hours. During this time bainite is formed. Bainite consists of ferrite plates between which, or inside which, short cementite rods have formed. Tensile strength and yield strength are increased without too large a decrease in ductility. Wear resistance is greater than that of other ductile irons and some steels, though not so good as white cast iron.

3.9.2 Uses of Various cast irons

The following are typical uses of the various cast irons;

(1) Grey iron - Water pipes, motor cylinders and pistons, machine castings, crankcases, machine tool beds, manhole covers.

(2) White iron - Water resistant parts, such as grindling mill parts, crusher equipment.

(3) Blackheart - Wheel hubs, pedals, levers, general hardware, brake shoes.

(4) Whiteheart - Wheel hubs, bicycle and motorcycle, frame fittings.

(5) Pearlitic malleable - Camshaft, gears, couplings, axle housings.

(6) Nodular ferritic - Heavy duty piping.

(7) Nodular pearlitic - Crankshafts

(8) Austempered ductile iron - Crankshaft, Camshafts, railway couplings, machinery guides, plough shares.

In selecting a cast iron for a particular application the following considerations are taken into account :

(1) Grey irons : Very good machinability and stability. Able to damp out vibrations, i.e. considered to have good damping capacity. Excellent wear resistance due to the graphite giving a self-lubricating effect for metal to metal contacts. Relatively poor tensile strength and ductility, but good strength in compression. Not good for shock loads.

(2) White iron : Excellent abrasion resistance. Very hard. Virtually unmachinable so has to be cast to the required shape and dimensions.

(3) Malleable iron : Good machinability and stability. Higher tensile strength and ductility than grey iron. Better for shock loads than grey iron.

(4) Nodular iron : High tensile strengths with reasonable ductility. Good machinability and wear characteristics, but not as good as grey iron. Better for shock loads than grey iron (about the same as malleable iron).

The ability to damp out vibrations, i.e. the damping capacity, referred to above is important as the higher the capacity the less noise there will be as a result of vibrations of a casting. The presence of graphite flakes or nodules enhances the damping capacity.

3.10 HEAT TREATMENT OF STEELS

The term heat treatment is used to describe a process involving controlled heating and cooling of a material in order to change its microstructure and properties. Vital factors in determining the outcome of heat treatment are the temperature to which the material is heated and the rate at which it is cooled back to room temperature.

All heat treatment processes may be considered to consist of three main parts.

(1) the heating of the metal to the pre-determined heat treating temperature,

(2) the soaking of the metal at that temperature until the structure becomes uniform throughout the section.

(3) the cooling of the metal at some pre-determined rate such as will cause the formation of, or will maintain desirable structures within the metal.

3.10.1. Reasons for heat treatments

Metal is heat treated to give it certain desired properties. Some of the properties which may be required and the treatments necessary are as follows:

To soften the metal - **Annealing**

To harden it to resist wear, or to enable it to cut other metals - **Hardening.**

To remove some of the extreme brittleness caused by hardening - **Tempering.**

To refine the structure after it has been distorted by hammering or working when in the cold state- **Normalising.**

The following is a brief overview of hardening, annealing and tempering.

3.10.2 The hardening steel - The martensite reaction hardening

Heating to produce a fully austenitic structure followed by rapid cooling, i.e. quenching, to produce a martenistic structure. Slow cooling from the austenitic structure, in which the carbon in the steel is dispersed throughout the structure, involves carbon atoms diffusing out of the austenitic structure to form cementite and hence pearlite, a laminar structure of cementite and ferrite. If, however, the steel is rapidly cooled, i.e. quenched by, for example, immersion in water, then there is insufficient time for the carbon atoms to diffuse out of the austenitic structure. They become trapped within the structure and the result is that the austenitic structure transforms to a new structure, a martenstic structure. This is not an equilibrium structure, hence it does not appear on the thermal equilibrium diagram. **The effect of martensite on the properties of the steel is to make it harder and more brittle.**

(a) The effect of carbon

The amount of carbon present directly determines the hardness of the martensite, is shown in Fig. 3.14.

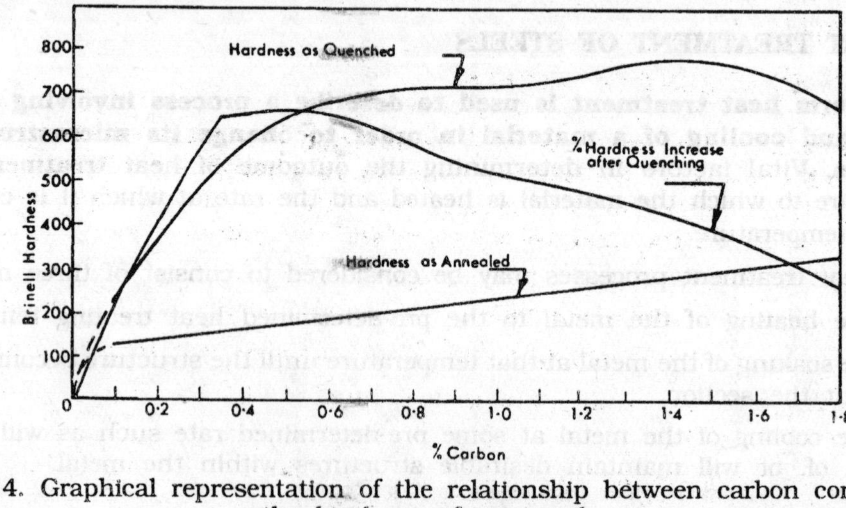

Fig. 3.14. Graphical representation of the relationship between carbon content and the hardness of martensile.

From this figure it is readily seen that the hardness of quenched plain carbon steel increases rapidly until the eutectoid composition is reached, when the hardness curve levels off and then increases slowly. Martensite formed in eutectoid steels is the hardest form possible and that in hypereutectoid steels has the same hardness. The slight increase in overall hardness in hyper-eutectoid steels is due simply to the presence of free carbide particles (Cementite in plain steels) which are themselves extremely hard and brittle.

(b) The effects of the rate of cooling

In order to understand the results of different cooling rates upon the structure and properties of austenitised and properties of austenitised steel it is interesting to examine four different samples of eutectoid steel that have been cooled at different rates (Fig. 3.15).

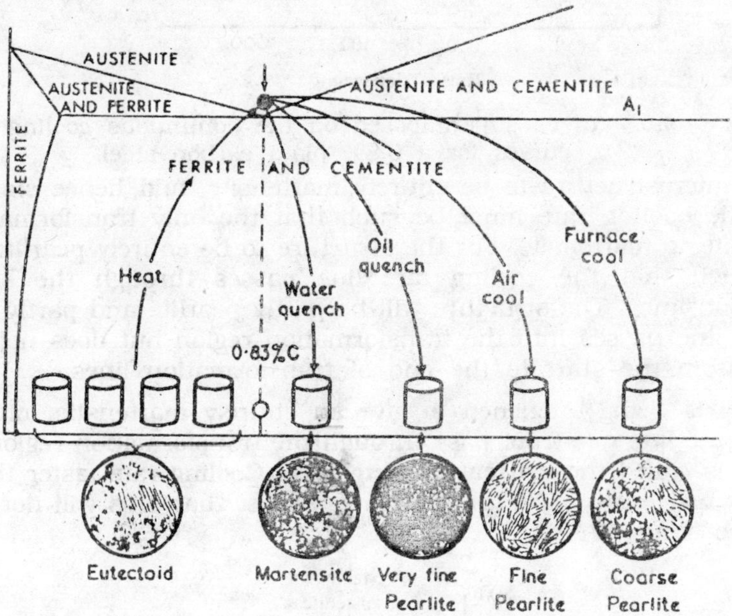

Fig. 3.15. Microstructures resulting from different cooling rate applied to austenitised samples of eutectoid steel.

Fig. 3.16 shows the continuous cooling transformation curve for a 0.8% carbon steel, i.e. the eutectoid composition. Superimposed on the diagram are graphs of temperature against time that could occur for the steel when quenched in different media, i.e. different cooling rate graphs. Which fast cooling (curve 1) from the stable austenite temperature the only transformation that occurs is from austenite to martensite. With slightly slower cooling rate (curve 2) a partial transformation to pearlite occurs and the final microstructure is thus martensite and pearlite. With an even slower cooling rate (curve 3) the austenite changes entirely to pearlite and the final structure is fine pearlite, no martensite being produced. With the yet slower cooling rate (curve 4) the austenite change entirely to pearlite again no martensite is produced.

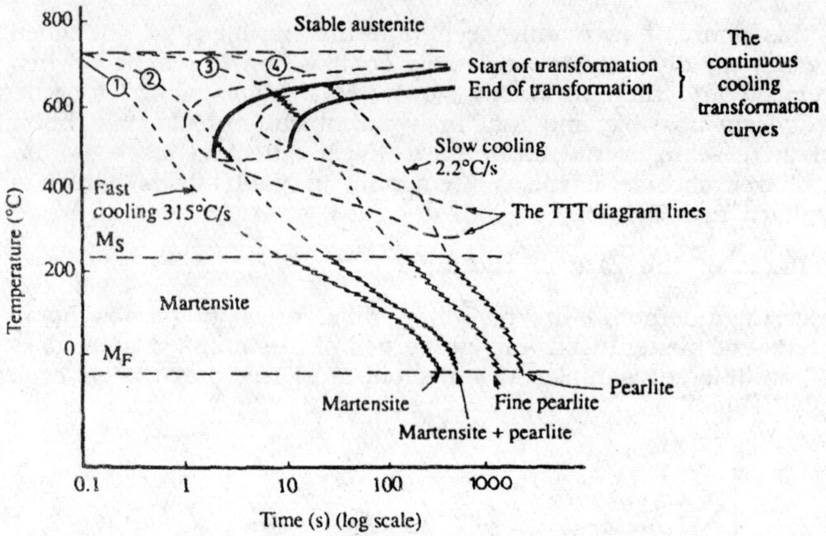

Fig. 3.16. Cooling curves superimposed on the continuous cooling transformation curves for a 0.8% plain carbon steel.

For the microstructure to be entirely martensite, and hence show the maximum hardness, the cooling rate must be such that the only transformation occurring is from austenite to martensite. For the structure to be entirely pearlite the cooling rate must be such that the cooling rate line passes through the start and end of transformation lines. The structure will be partly pearlite and partly martensite if the cooling rate line passes into the transformation region but does not pass completely through it from the start to the end of transformation lines.

If a steel is being quenched to give an entirely martenstic microstructure then the cooling rate line must not pass through the transformation region. This minimum cooling rate is called the **critical cooling rate**. Cooling rate faster than this will give a completely martensite structure but rates slower than this will not. Fig. 3.17 shows the critical cooling rate.

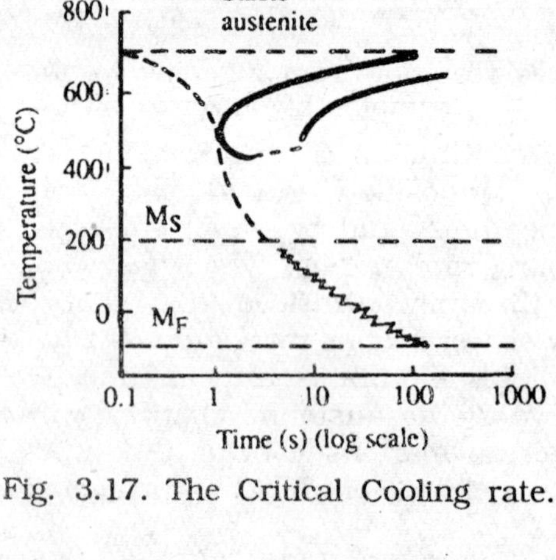

Fig. 3.17. The Critical Cooling rate.

(C) Effects of different quenching media

Different quenching media remove the heat from steel at different rates so that correct selection of a quenching medium for a particular grade of steel is of the utmost importance. **The cheapest quenching medium is plain water, but agitation must be used since there is a tendency for air bubbles to form on the surface of the article being quenched, these having an insulating effect and hence producing "soft spots". Salt water or brine is a more severe quenching medium than water, its main disadvantage being that it causes rapid rusting unless carefully washed away after quenching is complete.** For very low carbon steels hydroxide solutions are often used instead of brine since the quench is even more severe. **If slower cooling rates are desirable, various grades of oil with high flash points may be used, the grade determining the severity of the quench. Oil would always be used for high carbon steels since water quenching is too severe and results in cracking.**

(d) Hardness and hardenability

Steels possessing hardness and strength are required for structural uses and for applications to resist wear and distortion. The distinction should be made between hardness and hardenability. **Hardness is a measure of resistance to plastic deformation and hardenability is the ease with which hardness can be attained.** The maximum hardness of a steel is obtained only when it is heat treated to produce a structure of 100% martensite [Fig. 3.18]. A steel which rapidly transforms from

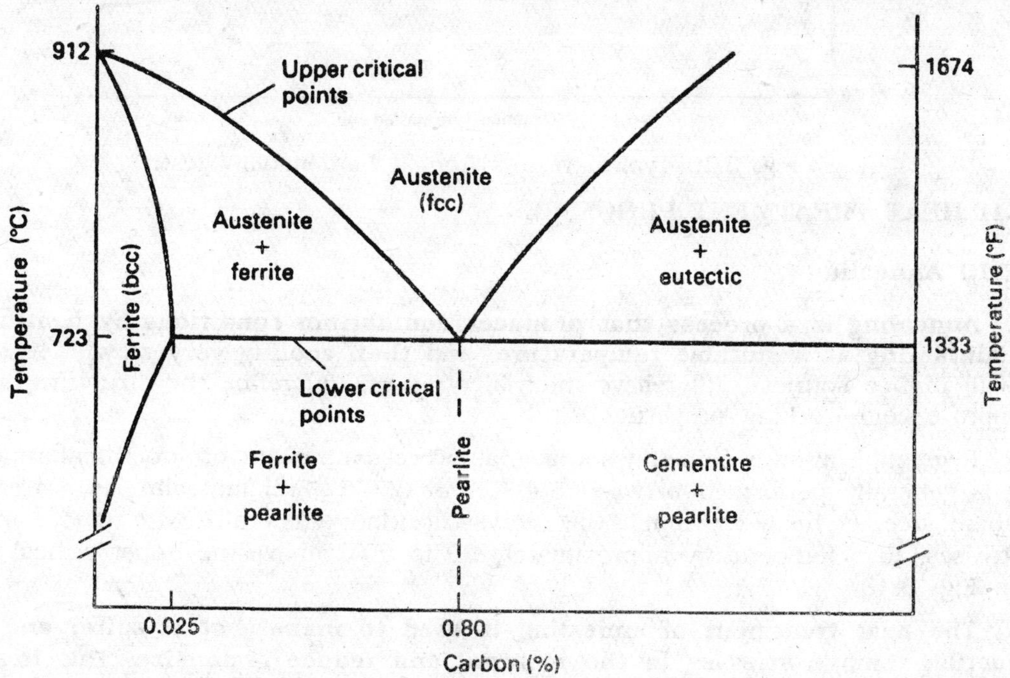

Fig. 3.18 Section of the iron-carbon diagram.

austenite into ferrite and carbide, rather than martensite, has a low hardenability. The depth of hardening of a steel is a measure of the hardenability. In plain carbon steels a hard surface layer of martensite can be formed by rapid cooling, below the surface ferrite and pearlite are formed due to the slower cooling rate. Alloy steel have lower critical cooling velocities and therefore greater hardenabilities than plain carbon steels. Martensite may be present at distances up to 50 mm below the surface. The Jominy test is often used to assess the hardenability of a steel. A series of hardeness measurements is then made along the length of the bar and a hardenability curve is obtained, as shown in Fig. 3.19. This curve will be obtained for all plain carbon and low alloy steels.

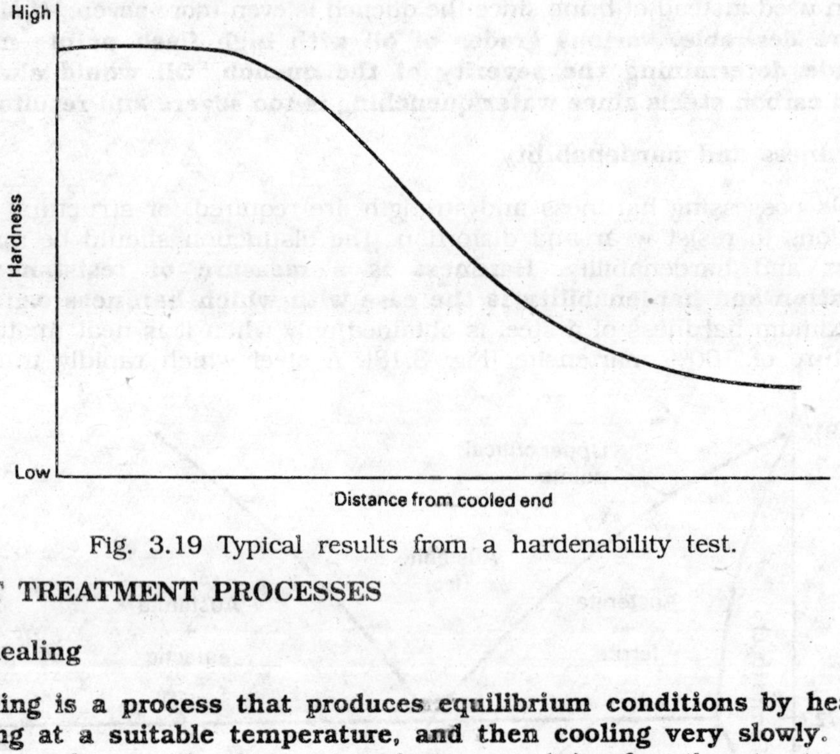

Fig. 3.19 Typical results from a hardenability test.

3.11 HEAT TREATMENT PROCESSES

3.11.1 Annealing

Annealing is a process that produces equilibrium conditions by heating and maintaining at a suitable temperature, and then cooling very slowly. It is used to (i) induce softness, (ii) relieve internal stresses, (iii) refine the structure and (iv) improve cold working properties.

Recrystallization is sometimes known as process annealing or subcritical annealing, it is generally performed between 500°C and 650°C. Full annealing is a treatment where steel is heated through the recrystallization temperature to remove internal stresses, to a temperature approximately 30 to 50°C above the upper critical points on Fig. 3.18.

The heat treatment of annealing is used to make a steel softer and more ductile, remove stresses in the material and reduce grain size. This treatment involves heating a steel to a temperature high enough to convert the structure of the steel to austenite, holding it there long enough for change to fully occur, then very slowly cooling to room temperature. This change has to be slow enough for the structure to be converted to ferrite and pearlite. The end result is a steel in as soft a condition as possible. The above annealing process is referred to as full annealing.

The austenizing temperature used is generally about 40°C above the upper critical temperature i.e. A_3 temperature. Fig. 3.20 shows the relevant austenizing temperature region on the iron- carbon diagram and Fig. 3.20, the temperature time relationship for the process with reference to the continuous cooling transformation curve.

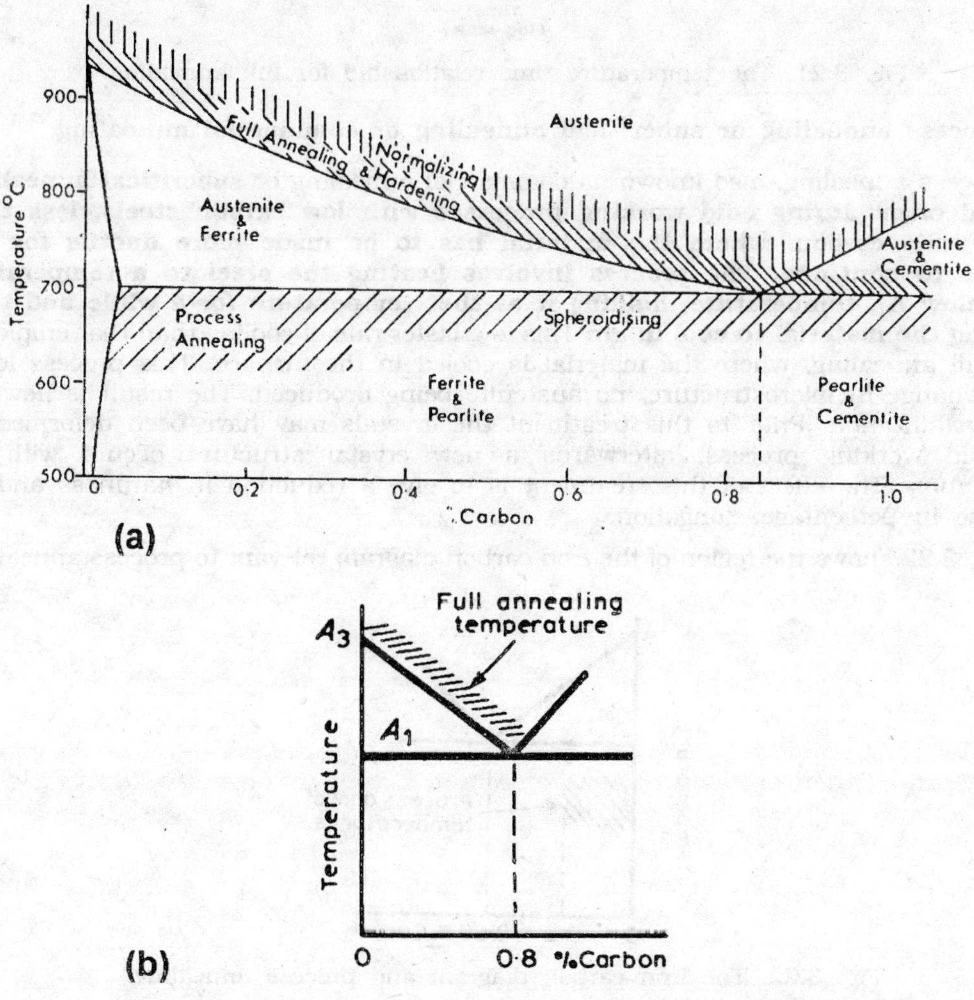

Fig. 3.20.(a) The iron carbon diagram and the temperature ranges shown on steel portion. (b) line diagram showing full annealing.

Full annealing is a slow and costly process and is not used as often as other softening processes since the resulting softness and less of strength is often both unnecessary and undesirable.

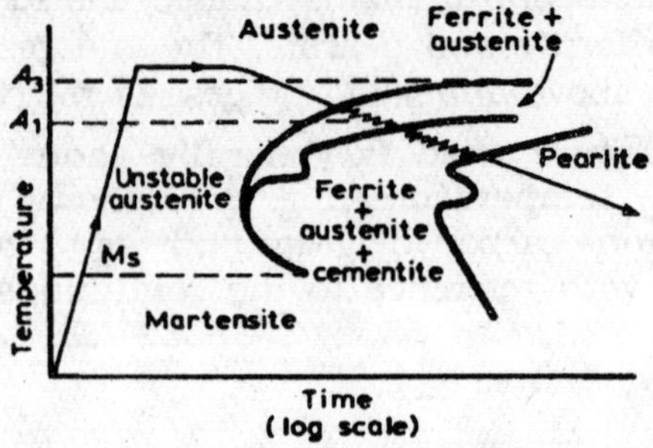

Fig. 3.21. The temperature time relationship for full annealing.

(b) Process annealing or subcritical annealing or commercial annealing

Process annealing, also known as commercial annealing or **subcritical annealing, is used often during cold working processes with low carbon steels, less than about 0.3% carbon, where the material has to be made more ductile for the process to continue. The process involves heating the steel to a temperature just below A_1, temperature, holding it at that temperature for a while and then allowing the material to cool in air.** This is a faster rate of cooling than that employed with full annealing, where the material is cooled in the furnace. This process leads to no change in microstructure, no austenite being produced. The result is however a recrystallization. Prior to this treatment the crystals may have been deformed by the cold working process, afterwards a new crystal structure occurs with no deformation. The effect of this treatment is to give a reduction in hardness and an increase in percentage elongation.

Fig. 3.22 shows the region of the iron carbon diagram relevant to process annealing.

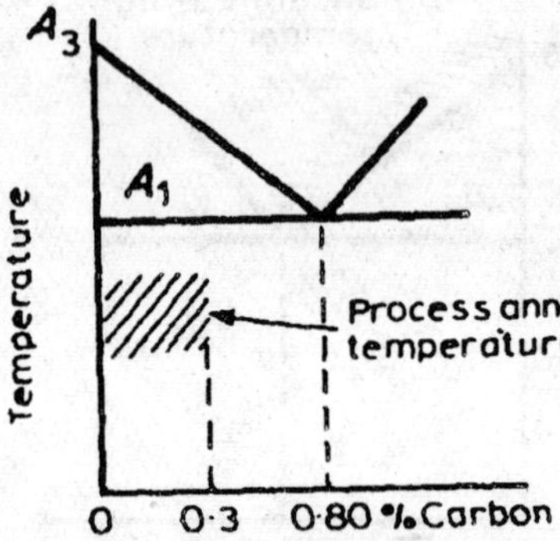

Fig. 3.22. The iron carbon diagram and process annealing.

When subcritical annealing is applied to steels having percentage of carbon higher than 0.3%, the effect of the heating is to cause the cementite to assume spherical shapes. Because of this the process is referred to as spheroidizing annealing. Fig. 3.23 shows the relevant part of the iron-carbon diagram for this treatment. The process results in a reduction in strength and hardness but an increase

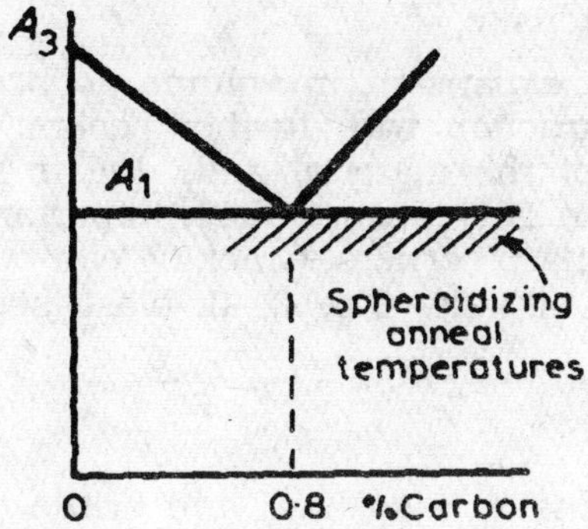

Fig. 3.23. The iron carbon diagram and spheroidizing annealing.

in ductility and machinability. The spherical pieces of cementite act as chip breakers during machining and so improve machinability.

3.11.2 Normalizing

Full annealing is an expensive process and normalizing is often used as an alternative where some loss in ductility is acceptable.

The term normalizing is used to describe an annealing process which involves heating a steel to about 40°C above A_3 temperature and then allowing the steel to cool in air. The process is thus similar to full annealing in that the microstructure is transformed to austenite but differs in that a faster rate of cooling is used (Fig. 3.24). The result of such a treatment is a ferrite and pearlite microstructure, the hardness and strength being slighting greater than that which would have occurred with full annealing.

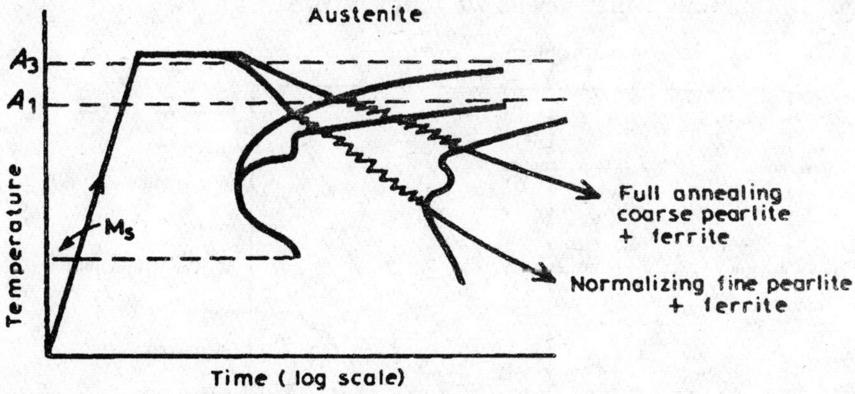

Fig. 3.24. The temperaure time relationship for full annealing and normalizing.

3.11.3 Quenching

Quenching involves cooling a steel very rapidly by immersion in oil or water. If an austenitic steel with more than 0.35% carbon is quenched, excess carbon is not precipitated but is trapped within the bcc ferrite lattice to form a supersaturated solid solution known as martensite.

As steel cools there is an expansion in volume as bcc austenite transforms to fcc ferrite, followed by a contraction with further cooling of the ferrite. This effect progresses towards the core of the material and, due to the brittleness caused by quenching, crancking can occur. This is particularly important for thin sections, abrupt changes in cross section and sharp corners. Cracking can be avoided by correct choice of the quenching medium, e.g. oil (alloy steels), air blast (special alloy steels), or water (plain carbon steel).

3.11.4 Tempering

After quenching, the martensite is so hard and brittle that it is of little practical use. To relieve the internal stresses brought on by quenching, reduce the hardness and improve both the toughness and ductlity, steel is subjected to a process called tempering. **Tempering is gentle reheating of the steel to between 180°C and 650°C,** the temperature depending upon the required properties (i.e. below the lower critical point). The carbon can then be precipitated in a more stable form. **Internal stresses are removed and the ductility and toughness are improved at the expense of the hardness and strength a compromise based upon the required properties.** The martensite is transformed into either sorbite by tempering above 450°C or troosite by tempering below 450°C. Troosite and sorbite are fine dispersions of iron carbide in a ferrite structure.

Sometimes quenching and tempering are performed as one process by rapid cooling to the tempering temperature, then holding to allow carbon precipitation. This reduces the risk of cracking due to the volume expansion when austenite transforms to ferrite. The carbide is then present in a different finely divided form known as bainite.

Fig. 3.25 shows schematically the various micrographic structural changes possible for tempering at different range levels of temperature.

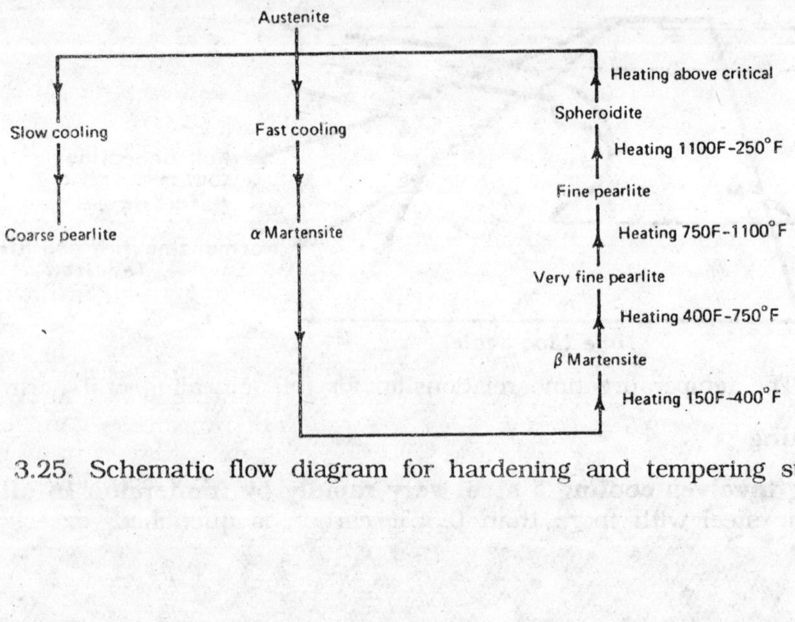

Fig. 3.25. Schematic flow diagram for hardening and tempering steel.

3.11.5 Surface hardening

A steel can be directly hardened by quenching and austenitic structure. The hardness and properties of the steel produced depend upon the carbon content, the heating time, the temperatures used and the cooling rate. As the hardness and strength of a material increase, then the toughness decreases. Many applications require a tough core of material with a very hard surface, which can be achieved by heat treating a steel to produce the required hardness only at the surface. **Surface hardening methods can involve heating either the whole component (e.g. case hardening and nitriding) or only the surface of the component (e.g. flame or induction hardening). All methods except induction hardening involve diffusion of another material into the metal surface.**

3.11.6 Case hardening

This is a process for increasing the surface hardness of a steel, and requires heating of the whole component. This process is usually performed on low carbon steels (approximately 0.15% carbon) which will not respond to direct hardening.

The **first stage** involves heating the steel in contact with carbon, to increase the carbon content at the surface. This is known as **carburizing or case carburizing.** The steel is heated above its upper critical point (approximately 908°C) and is surrounded either by carbon in solid form (**pack carburizing**), or by liquid sodium carbonate mixed with either sodium or barium chloride. Alternative **gas carburizing** can be used with an atmosphere of methane or mixed hydrocarbon gases. The temperature must be maintained long enough for the carbon to penetrate to the required case depth, usually 0.5 mm to 0.7 mm. The metal surface is transformed from a low carbon to a high carbon steel.

During carburizing, grain growth occurs within the steel. The **second stage** of the process involves heat treatment to harden the high carbon skin. The carburizing process causes a variation in carbon content (approximately 0.83 - 0.15% carbon) from the case to the core over a depth of about 0.5 mm. The difference between these regions is sometimes taken to be the point where the hardness measures Rockwell C50. The heat treatment required by these two regions is different and requires different temperatures. The core structure is first refined by heating above the upper critical point of the core composition (approximately 870°C), it is then quenched to prevent grain growth during cooling. This high temperature produces a coarse austenitic structure within the case, and when this is quenched it transforms to extremely brittle martensite. The case is then refined, by heating to just above the lower critical point (approximately 760°C) of the higher carbon layer; this also tempers the outer layers of the core. The steel is then quenched and final tempering of the case is carried out at 200°C to relieve quenching stresses. The minimum of hardened surface metal (less than 20%) should be removed after carburizing as it detracts from the purpose of this heat treatment process.

Carbonitriding : Carbonitriding is liquid carburizing with diffusion of some nitrogen. High hardness can be achieved with the following advantages: hard and shallow case, lower temperatures and less distortion. The process can be applied to any steel that can be carburizied, and also for materials that cannot be nitrided. It is a useful way of providing a wear resistant surface on cheap low-carbon steels.

Other carburizing processes have been developed such as **Vacuum carburizing.** Parts of the cycle are carried out under vacuum and process times are reduced, although operating costs are higher. Carburizing can also be followed by other heat treatments to provide particular advantages. However, any additional processing costs must be justified.

Cyaniding is a similar to carburizing but the steel component is heated in a bath of molten sodium cyanide and sodium carbonate at 950°C. Both carbon and nitrogen diffuse into the steel surface and the surface hardness is increased by the presence of iron nitrides. The component must still be heat treated. Components must usually be ground to the final dimensions, due to volume changes during the case-hardening process.

3.11.7 Nitriding

This process can only be used for low-alloy steels (that can be directly hardened) which contain small amounts of chromium, nickel, vanadium or aluminum. The carbon content is usually between 0.2% and 0.5% carbon. Plain-carbon steels cannot be hardened by this process because the nitrides would form throughout the structure causing brittleness. It is also used for same tool steels, gray cast irons, medium carbon steels (400 series) and special alloy steels (high aluminium or chromium). **The material is first heat treated and oil quenched to strengthen the core by providing required hardness and toughness, and then tempered at 550°C to 750°C to relieve stresses. It is machined to allow for the small growth of 0.02 mm during nitriding.**

The component is then nitrided by heating to 500°C in a constant closed circulation of ammonia. A case depth of 0.1 mm can be produced in about 10 hours, although a depth of 0.7 mm can take up to 100 hours. The core properties are unaffected by nitriding. Finally, the temperature is allowed to fall to 150°C before the component is removed from the ammonia atmosphere.

The advantages of nitriding are that an extremely hard surface is produced, cracking and distortion are minimized due to the lower temperatures used, and no subsequent heat treatment is required. The component can be produced more cheaply because machining can be performed before hardening, due to the small volume changes which occur.

The disadvantages of nitriding are the costs of the surface pretreatments and the length of the nitriding cycles required. The range of materials for which nitriding is applicable is small and same of these materials are expensive and difficult to machine. The nitrided case produced is relatively thin and only a small amount of material can be subsequently removed by machining. However, nitriding is an accurate process and machining should not be necessary. Masking paints or plating (both subsequently removed) can be applied to enabled nitriding to occur only in selected regions.

Soft nitriding is gas nitriding with carbon added to the furnace atmosphere. Extremely precise case depths can be produced which are shallower than for the conventional process, and the applicable range of alloys is wider.

3.11.8 Flame hardening and induction hardening

In both these processes the steel (containing between 0.4% and 0.7% carbon) is rapidly heated to the hardening temperature, using an oxy-acetylene flame or a high frequency induction coil. The surface is rapidly quenched to form martensite, the subsequent cooling and outflow of heat from the core tempers the surface layers. **The flame temperature must be above the melting point of the steel, surface melting can occur if case hardening depths beyond 0.8 mm are attempted. For induction hardening, it takes about 5 seconds for a case depth of 3 mm to reach its hardening temperature.**

The time that the component is heated is the crucial factor. After the required depth of metal is heated, the part must be quenched immediately. Problems can occur if the cross section changes abruptly, or if there are thin walled sections that may heat all the way through. Induction hardening is faster, requires less expensive equipment, and can harden thicker cases than carburizing. Cracking is the main problem, especially with high alloy steels that are more sensitive heat transfer into the core. A component may be case hardened at selected regions to reduce wear and abrasion, and nonhardened areas are easier to machine. However, stresses will be present in the hardened/soft metal region, and these should be located away from features which induce stress concentrations.

3.11.9 Age hardening or precipitation hardening

The change in a metal by which its structure recovers from an unstable condition produced by quenching (quench aging) or by cold working (strain aging). The change in structure is due to the precipitation of one of the constituents from a saturated solid solution and results in a material that is stronger and harder but usually less ductile. This type of aging takes place slowly at room temperature; it may be accelerated by a slight increase in temperature. The process is then called precipitation hardening (also stress relieving).

Age hardening occurs when the solubility of an alloying element decreases as the temperature decreases. At lower temperatures the element in solution becomes supersaturated and starts to precipitate out, causing increased hardness and improved mechanical properties. This age hardening process develops with time. Particular alloying elements are added to the steel and they precipitate out of the structure at lower temperatures, thus improving the strength and hardness. Maintaining the material at the aging temperature for a prolonged time can cause softening following the aging process, known overaging.

3.11.10 Austempering and marquenching

Austempering involves quenching an austenitic structure to a temperature above that at which martensite is formed. The steel is held at that temperature (isothermal transformation) until bainite formation is complete. Bainite is a more ductile phase than martensite. An austempered steel does not require tempering as it is not stressed and does not suffer from distortion.

Marquenching is sometimes called interrupted quenching or martempering, it is used to harden steel without cracking sometimes caused by quenching. An austenitic steel is quenched to a temperature above that required for martensite transformation; this temperature is maintained in order to establish thermal equilibrium throughout the entire component. Slow cooling then produces tempered martensite, although this may be followed by further tempering.

Austempering and marquenching are both lengthy and costly operations.

However, their advantages are the production of less stressed parts and less risk of cracking or distortion than when quenching is performed. Both processes can be readily used on high-alloy steels, but plain carbon steels require more control if a thick section is being treated. Sometimes, with rapid quenching, small quantities of austenite do not have sufficient time to transform to martensite. This austenite is trapped in the microstructure, but it transforms later (several hours or days) causing dimensional changes. This effect can occur with quenching and tempering, or marquenching.

3.12 HEAT TREATMENT FOR ALLOY STEEL

A relatively new class of steels known as **maraging steels** has been developed. These steels are strong and hard, but less brittle than conventional heat hardened steels. A steel containing the minimum amount of carbon is alloyed with between 18% and 25% nickel and other alloying elements. A typical composition is 18% nickel, 8% cobalt, 5% molybdenum and 0.4% titanium. This alloy has a type of martenistic structure even when slowly cooled, and it is hard and tough. The other alloying elements improve the strength. The steel is annealed and martensite is formed; this is followed by shaping and machining. Hardening is achieved by heating at 450°C to 500°C for three hours, followed by air cooling. The surface hardness can be improved at this stage by nitriding. The lower temperature treatment means that cracking and distortion are eliminated. Rapid heating does not cause hardening, which makes these steels suitable for welding. However, maraging steels (named from martenistic age hardening) are more costly to produce.

The addition of nickel to steel that have been carburizing prevents grain growth, and chromium is added to offset the mass effect in thicker components. Typical nickel case-hardened steels contain 0.12% carbon, 3% nickel, 0.45% manganese and are water quenched. The low carbon content means that the steel does not respond to direct hardening. If the nickel content is increased to 5%, then the steel may be oil quenched. A steel containing 15% carbon, 4% nickel, 0.8% chromium and 0.4% manganese has improved strength and hardness due to oil quenching.

Steels containing sufficient chromium (about 2%) can be quenched in air. Alloying elements are added to steel so that the critical cooling rate and the hardening temperature are reduced. A steel containing 1.5% manganese (0.35% carbon) is relatively cheap and can be oil quenched to give good strength. A steel containing 0.3% carbon, 3% nickel, 0.6% manganese has good strength and hardness when oil quenched. The addition of chromium improves the toughness of a steel and lowers the hardening temperature.

Air quenching can be employed if sufficient elements are added, and this reduces the risk of distortion. Typical steels contain 0.3% carbon, 3% nickel, 0.8% chromium, 0.6% manganese for oil quenching and 0.3% carbon, 4.25% nickel, 1.25% chromium, 0.5% manganese for air cooling. When nickel-chromium steels are tempered between 250°C and 400°C, depending upon the composition, they may became brittle. This can be avoided by the addition of 0.3% molybdenum, which also reduces mass effect by forming carbides. About 0.5% vanadium improves the shock resistance of chromium steels, but can give rise to mass effect if it is used to replace nickel.

Table 3.6 presented a comparison of some of the important properties of the main types of irons and steels. The compositions quoted are approximate and the values of the tensile strength and yield strength are average values for the ranges normally encountered. This table is meant to provide a quick and easy comparison of different materials. It is not comprehensive : many other alloys and properties could have been included.

Table 3.6. Properties of ferrous metals and alloys.

Material and approximate composition (mass %)	Melting point or range, °C (°F)	Youngs modulus GN/m^2 (psi x 10^{-6})	Tensile strength MN/m^2 (psi x 10^{-3})	Yield strength MN/m^2 (psi x 10^{-3})	Coefficient of linear thermal expansion (x 10^6)
Iron; annealed (C<0.02; Mn<0.02)	1536 (2797)	200 (29.0)	290 (42.0)	130 (18.0)	11.8
Gray cast iron (C2.75-3.50; Si=2.0; Mn=0.7)	1120-1175 (2048-2147)	100 (14.5)	250 (36.2)	—	10.9
Malleable cast iron; ferritic (C=2.3; Si=1.0; Mn0.55 max.; P and S <0.20 each)	1120-1175 (2048-2147)	170 (24.6)	365 (52.9)	240 (34.8)	13.6
Low carbon steel; hot rolled (C0.1-0.25; Si 0.1; Mn = 0.6; P 0.04 max.)	1510-1525 (2750-2777)	205 (29.7)	400 (58.0)	250 (36.2)	15.3
Medium carbon steel;hot rolled (C 0.28 - 0.50; Si 0.1-0.6; Mn = 0.7; P and S 0.04 max. each)	1480-1510 (2696-2750)	205 (29.7)	600 (87.0)	375 (54.3)	15.1
High carbon steel; hot rolled (C0.5-1.05; Si 0.1-0.5; Mn = 0.6; P and S 0.04 max. each)	1425-1470 (2597-2678)	205 (29.7)	850 (123.2)	500 (72.5)	14.7
Ferritic stainless steel cold worked (C 0.12 max; Mn 1.00 max; Cr 14-18; Si 1.00 max; P and S 0.04 max. each	1425-1510 (2597-2750)	200 (29.0)	550 (79.7)	435 (63.0)	10.5
Martensitic stainless steel; heat treated (C 0.15 max; Cr 11.5-13.5; Mn 1.00 max.; Si 1.00 max; P 0.04 max.)	1480-1530 (2696-2786)	200 (29.0)	1000 (145.0)	700 (101.4)	10.0
Austenitic stainless steel; cold worked (C 0.08 max; Cr 16-18; Ni 10-14; Mo 2-3; Mn 2.0 max; Si 1.0 max.)	1370-1400 (2498-2552)	190 (27.5)	850 (123.2)	600 (87.0)	16.2
Stainless stee; precipitation hardening (C 0.09 max.; Cr 17.0; Ni 7.1; Al 1.25; Mn and Si 1.0)	—	200 (29.0)	1350 (195.7)	1280 (185.5)	15.5
Marginal steel; oil quenched & aged (C 0.026; Ni 18.5; Co 7.0; Mo 4.5; Mn 0.1; Si 0.11; Ti 0.22; B 0.003	—	180 (26.1)	1900 (275.4)	1850 (268.1)	10.2

3.13 NON-FERROUS ALLOYS

The term ferrous alloys is used for those alloys having iron as the base element. The term non ferrous alloys is used for those alloys which do not have iron as the base element.

3.13.1. Advantages of non-ferrous alloys over ferrous alloys

(1) Good resistance to corrosion without special processes having to be carried out.

(2) Most non ferrous alloys have a much lower density and hence lighter weight components can be produced.

(3) Casting is often easier because of the lower melting points.

(4) Cold working processes are often easier because of the greater ductility.

(5) Higher thermal and electrical conductivities.

(6) More decorative colours.

3.13.2. Advantages of ferrous alloys over non-ferrous alloys

(1) Generally great strengths.

(2) Generally great stiffness.

(3) Better for welding.

3.13.3 The range of non-ferrous alloys

The following are some of the **non ferrous alloys in common use in engineering.**

(1) Aluminium alloys : Aluminium alloys have a low density, good electrical and thermal conductivity, high corrosion resistance. Typical uses are metal boxes, cooking utensils, aircraft bodywork and parts.

(2) Copper alloys : Copper alloys have good electrical and thermal conductivity, high corrosion resistance. Typical uses are pump and valve parts, coins, instrument parts, springs, screws. The names brass and bronze are given to some forms of copper alloys.

(3) Magnesium alloys; (4) Nickel alloys; (5) Titanium alloys; (6) Zinc alloys.

3.14 ALUMINIUM AND ITS ALLOYS

Second to steel, aluminium is the most widely used metal and its applications are manifold. It is available in all forms, such as wire, bars, extrusions (including structural shapes), power, sheets, plates, forging, and castings. Aluminium possesses a high resistance to most corrosive atmospheres because it readily forms a passive oxide surface coating. Different alloying elements affect the corrosion resistance of aluminium to varying degrees, however, all alloys are essentially corrosion resistant. **Halogen acids and strong alkalies** are two of the few types of solutions that will attack aluminium because these substances can remove the oxide coating and expose fresh aluminium.

Aluminium has good electrical and thermal conductivity as well as high light reflectivity. It is easily formed, drawn, stamped, spun, machined (with exception of "O" temper), welded, or brazed, depending upon the temper of the particular alloy.

Alluminium alloys can be divided into two groups, wrought alloys and cast alloys, Each of these can be divided into two further groups :

(1) Those alloys which are not heat treatable.

(2) Those alloys which are heat treated.

The term wrought material is used for a material that is suitable for shaping by working processes, e.g. forging, extrusion, rolling. The term cast material is used for a material that is suitable for shaping by a casting process.

3.14.1. Non heat treatment wrought alloys

Common alloys in this category are aluminium with manganese or magnesium. **A common aluminium - manganese alloy has 1.25% manganese**, the effect of this manganese being to increase the tensile strength of the aluminium. The alloy still has a high ductility and good corrosion properties. This leads to uses such as kitchen utensils, tubing and corrugated sheet for building. **Aluminium - magnesium alloys have up to 7% magnesium.** The greater the percentage of magnesium, the greater the tensile strength. The alloy still has good ductility. It has excellent corrosion resistance and thus finds considerable use in marine environments, e.g. constructional materials for boats and ships.

3.14.2. Heat treatable wrought alloys

The heat treatable wrought alloys can have their properties changed by heat treatment. Copper, magnesium, zinc and silicon are common additions to aluminium to give such alloys.

Table 3.7 lists three typical heat treatable wrought alloys of these, **the duralumin has the best formability prior to heat treatment, the magnese-silicon alloy the best corrosion resistance and the copper-Magnese-Zinc alloy the greatest strength after heat treatment.**

Table 3.7 Three heat treatable wrought aluminium alloys and their uses

S.N.	Type of alloy	Composition	Uses
1	Duralumin	4% Cu, 0.8% Mg, 0.7% Mn and 0.5% Si	Widely used in aircraft body work
2	Magnesium Silicon alloy	1% Si, 0.7% Mg 0.6% Mn	Ladders, scaffold tubes, container bodies, structural members for road and rail vehicles
3	Copper-Magnesium-Zinc alloy	5.5% Zn, 2.8% Mg, 0.45% Cu, 0.5% Mn.	Structural applications in aircraft and space craft.

3.14.3. Age hardening in aluminium alloys

Fig. 3.26 shows the thermal equilibrium diagram for aluminium - copper alloys. When such an alloy, say 3% copper -97% aluminium, is slowly cooled, the structure at about 54°C is a solid solution of the phase. When the temperature falls below

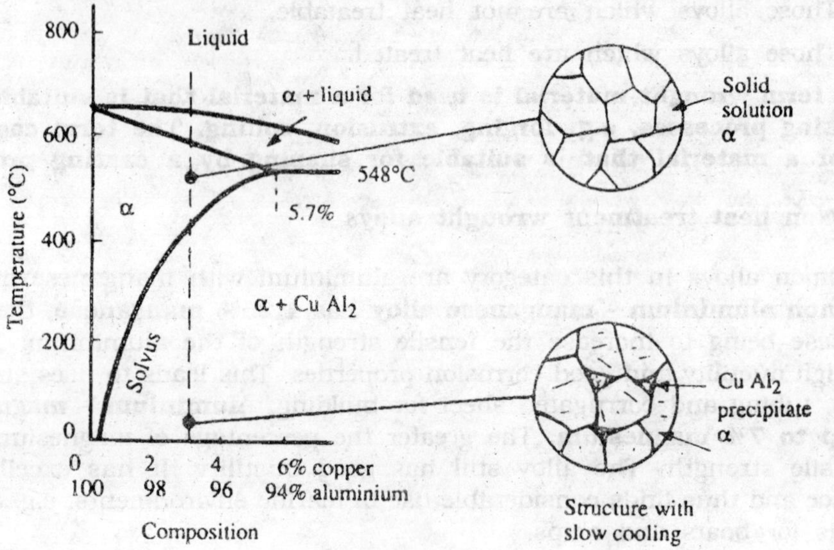

Fig. 3.26 Thermal equilibrium diagram for aluminium copper alloys

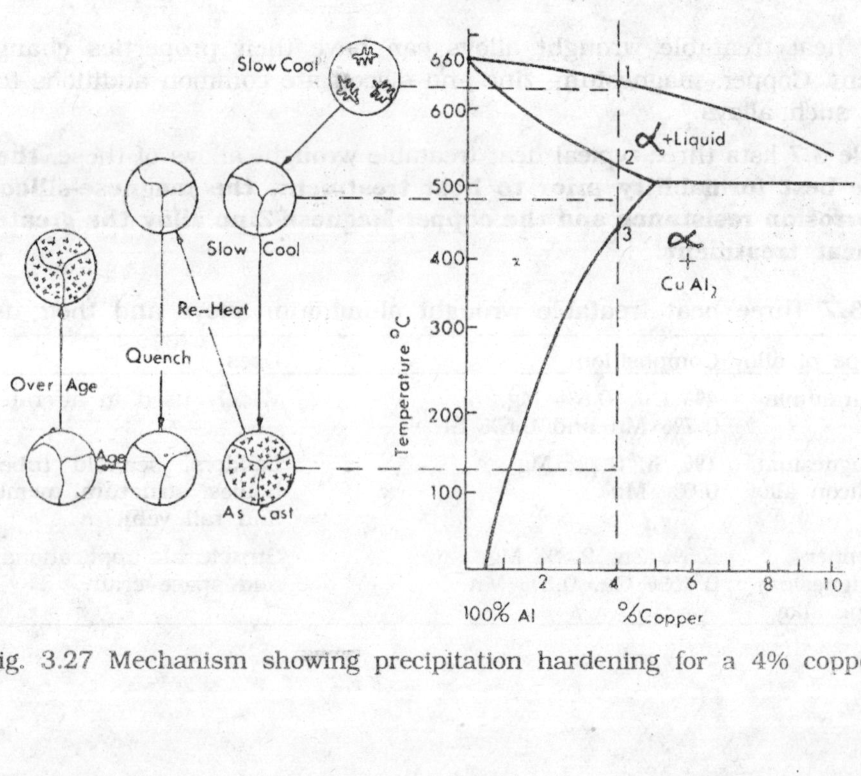

Fig. 3.27 Mechanism showing precipitation hardening for a 4% copper

the solvus temperature a copper - aluminium compound is precipitated. The result at room temperature is solid solution with this copper - aluminium compound precipitate ($CuAl_2$). The Precipitate is rather coarse, but this structure of the alloy can changed by heating to about 500°C soaking at that temperature, and then quenching, to give a supersaturated solid solution, just α phase with no precipitate. This treatment, known as **solution treatment,** results in an unstable situation. With time a fine precipitate will be produced. Heating to, say, 165°C for about ten hours hastens the production of this fine precipitate (Fig. 3.28). The microstructure with this fine precipitate is both stronger and harder. The treatment is referred to as **precipitation hardening**.

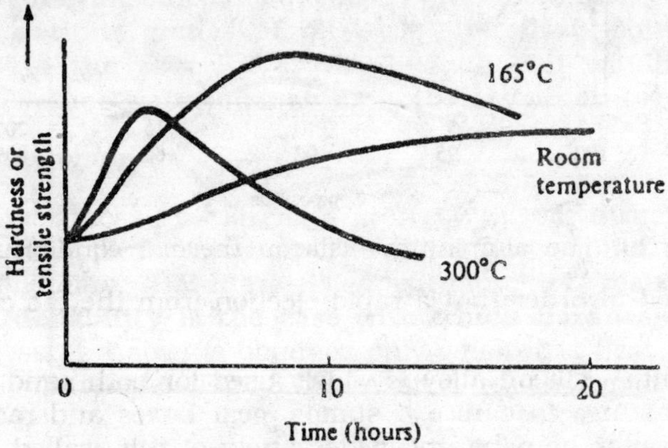

Fig 3.28 The effect of time and temperature on hardness and strength for an aluminium Copper alloy.

3.14.4. Non heat treatable cast alloys

An alloy for use in the casting process must flow readily to all parts of the mould and on solidifying it should not shrink too much and any shrinking should not result in fractures. In choosing an alloy for casting, the type of casting process being used needs to be taken into account. In **sand casting** the mould is made of sand bonded with clay or a resin. The cooling rate with such a method is relatively slow. A material for use by this method must give a material of suitable strength after a slow cooling process. With **die casting** the mould is made of metal and the hot metal is injected into the die under pressure. This results in a fast cooling. A material for use by this method must develop suitable strength after fast cooling.

A family of aluminium alloys that can be used in the 'as cast' condition, i.e. no heat treatment is used, has aluminium 9 and 13% silicon. These alloys can be used for both sand and die casting. Fig. 3.29 shows the thermal equilibrium diagram for aluminium - silicon alloys. The addition of silicon to aluminium increases its fluidity, between about 9 to 13% giving a suitable fluidity for casting purposes. The eutectic for aluminium - silicon alloys has a composition of 11.6% silicon. An alloy of this composition changes from the liquid to solid state without any change in temperature, and alloys close to this composition solidify over a small temperature range (Fig. 3.30). This makes them particularly suitable for die casting where a quick change from liquid

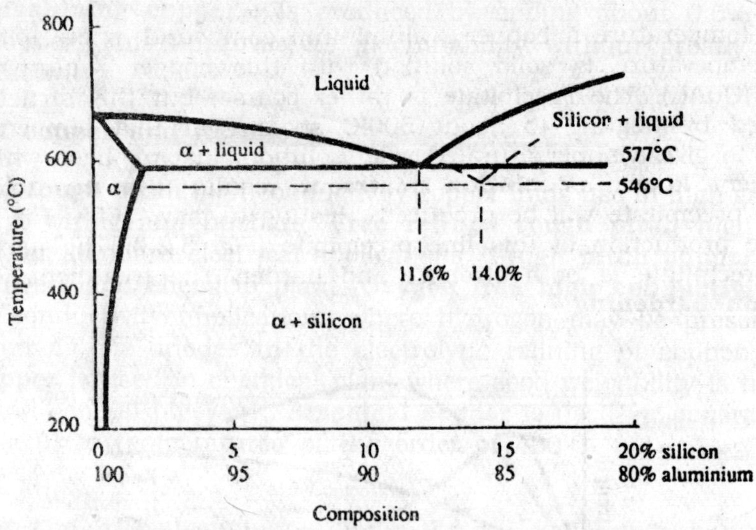

Fig. 3.29. The aluminium - silicon thermal equilibrium diagram.

to solid is required in order that a rapid ejection from the die can permit high output rate.

They aluminium - silicon alloy is widely used for both sand and die casting, being used for many castings in cars e.g. sumps, gear boxes and radiators. It is also used for pump parts motor housing and wide variety of this walled and complex castings.

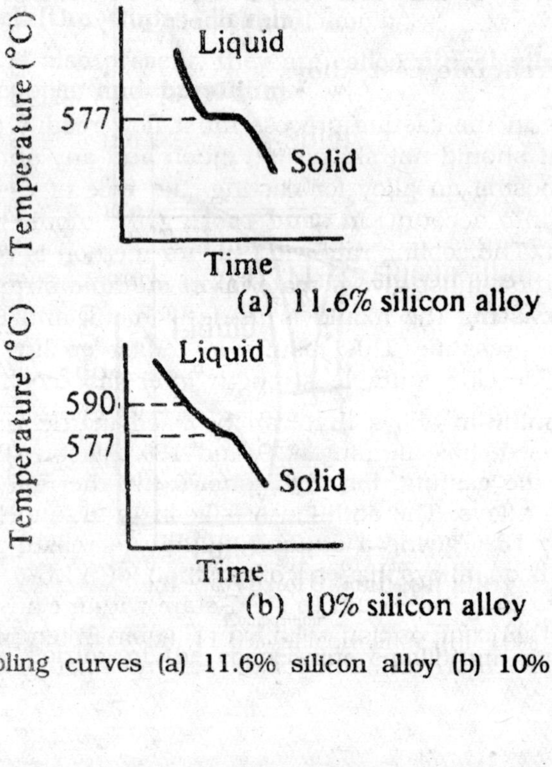

Fig. 3.30. Cooling curves (a) 11.6% silicon alloy (b) 10% silicon alloy.

Other cast alloys that are not heat treated are aluminium- silicon-copper alloys, e.g. 5% silicon and 3% copper, and aluminium - magnesium - manganese alloys, e.g. 4.5% magnesium and 0.5% manganese. The silicon - copper alloys can be both sand and die cast, the magnesium - manganese alloys are however only suitable for sand casting. They have excellent corrosion resistance and are often used in marine environments.

3.14.5. Heat treatable cast alloys

The addition of copper magnesium and other elements to aluminium alloys, either singly or in some suitable combination, can enable the alloy to be heat treated. Thus an alloy having 5.5% silicon and 0.6% magnesium can be subjected to solution treatment followed by precipitation hardening to give a high strength casting material. Another heat treatable casting alloy has 4% copper, 2% nickel and 1.5 magnesium.

3.15 COPPER AND ITS ALLOYS

Copper has very high electrical conductivity (second only to silver) and thermal conductivity can be manipulated readily by either hot or cold working. **Pure copper is very ductile and relatively weak.** Copper has good corrosion resistance. This is because there is a surface reaction between copper and oxygen in the air which results in the formation of thin protective oxide layer. The following grades of copper are commercially available.

(1) Tough pitch copper : Electrolytic tough pitch high conductivity copper is produced from cathode copper which has been melted and cast into billets, and other suitable shapes, for working. It contains a small amount of oxygen, present in the form of cuprous oxide, which has little effect on the electrical conductivity of the copper. If this grade of copper is heated to about 400°C in an atmosphere "gassing" occurs due to a reaction between cuprous oxide impurily and hydrogen or carbon monoxide which penetrates into the metal during heating. This reaction produces steam which forces copper grains slightly apart, the intercrystallive grades so produced severely reducing ductility. For this reason tough pitch copper is not used when welding or brazing is involved.

(2) Oxygen free high conductivity copper (OFHC Copper) : is made by remelting and casting tough pitch copper either in vacuo or in a non-oxidising atmosphere. It therefore contains neither dissolved oxygen nor residual deoxidant, and is the preferred material for all current carrying members such as wires and cables, busbars and terminals.

(3) Phosphorus deoxidized copper : is formed when electrolytic copper is remelted and deoxidized with about 0.5% phosphorus, the phosphorus having a much higher affinity for dissolved oxygen than the copper itself. **This grade is most suitable for gas welding and brazing applications.** However, the residual 0.05% or so of phosphorus in the copper severely reduces its electrical conductivity.

(4) Arsenical copper : The addition of about 0.5% arsenic to copper increases its tensile strength, especially at temperatures of about 400°C. It also improves its corrosion resistance but greatly reduces the electrical and thermal conductivities. This type of copper is known as arsenical copper.

(5) Free machining copper : is produced by adding about 0.5% tellurium or selenium to copper. This improves its machinability without greatly reducing its electrical conductivity.

3.15.1. Uses of copper and its alloys

Electrolytic tough pitch high conductivity copper finds use in high grade electrical applications e.g. wiring and busbars. **Free refined tough pitch** high conductivity copper is used for standard electrical applications. **Tough pitch copper** is also used for heat exchangers and chemical plant. **Oxygen free high conductivity copper** is used for high conductivity applications where hydrogen may be present, electronic components and as the anodes in the electrolytic refining of copper. **Phosphorus deoxidized copper** is used in chemical plant where good weldability is necessary and for plumbing and general pipework. **Arsenical copper** is used for general engineering work, being useful to temperatures of the order of 400°C.

3.15.2. Copper alloys

Copper may be alloyed with a wide range of other elements to produce many different alloy groups of industrial importance. The most important are :

(a) Copper - Zinc (the Brasses)

(b) Copper - Tin (the Tin Bronzes)

(c) Copper - Tin - Phosphorus (the Phosphorus- Bronzes)

(d) Copper - Aluminium (the Aluminium Bronzes)

(e) Copper - Nickel (the Cupro-Nickels)

Though where zinc is also present, they are called **nickel silvers**. A less common copper alloy involves copper and **beryllium**.

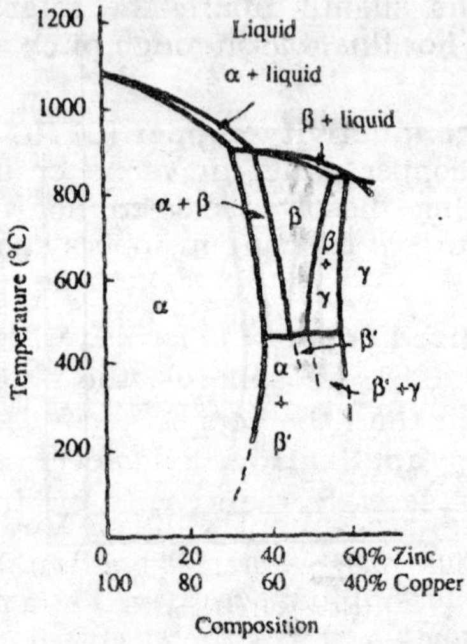

Fig. 3.31. Portion of the copper-zinc equilibrium diagram.

(a) Brasses : The brasses are copper zinc alloys containing up to 43% zinc. Fig. 3.31 show the relevant part of the thermal equilibrium diagram.

Three phases are present in the portion of the copper-zinc equilibrium diagram.

(i) The alpha (α) phase : a solid solution of up to 36% zinc in copper, which is malleable, ductile and hence readily cold worked.

(ii) The beta (β) phase : another solid solution, characteristically hard and brittle; it can, however, be hot worked very readily.

(iii) The gamma (γ) phase : also a hard and brittle solid solution, differening from the beta phase in that it cannot be hot worked. Thus, very few commercial brasses contain more than about 52% zinc, this being the limit above which the gamma phase appears.

The following copper-zinc alloys are of commercial importance

(1) 70/30 Brass or cartridge Brass : Cartridge brass, 30% zinc and frequently referred to as 70/30 brass, is used where high ductility is required with relatively high strength. It is called cartridge brass because of its use in the production of cartridge and shell cases.

(2) Gilding brass : Gilding brass, 15% zinc, is used for jewellery because it has a colour remembling that of gold and can so easily be worked.

(3) Alpha-beta (α – β) or duplex brasses - Between 900°C and 453°C the two phases are α and β. At 453°C this β phase transforms to a low temperature modification referred to as β phase. Thus at room temperature the two phases present are α and β′. The presence of the β phase produces a drop in ductility but an increase in tensile strength to the maximum value for a brass (Fig. 3.32). **These brasses are known as alpha-beta or duplex brasses. They are not cold worked but have good properties for hot farming processes, e.g. extrusion.** This is because the β phase is more ductile than the β phase and hence the combination of α plus β gives a very ductile material.

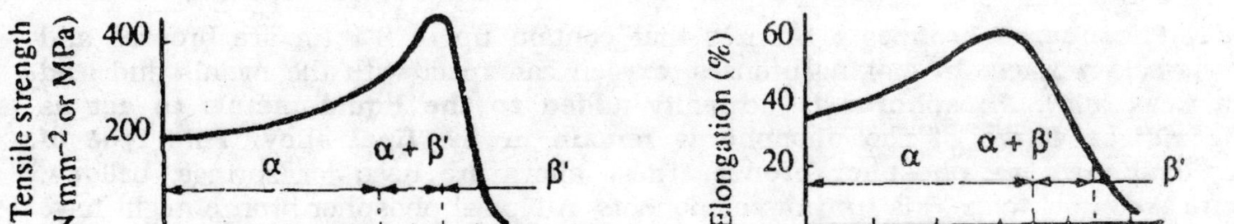

Fig. 3.32. Strength and ductility for copper-zinc alloys.

(4) 60/40 Brass or Muntz metal : This is an α-β brass and is suitable only for hot working applications. It has higher strength than the 70/30 brass and is commonly used to make hot rolled products valved stems, condenser tubes, and the like.

The hot working should take place at temperature in excess of 453°C. **The name Muntz metal is given to brass with 60% copper and 40% zinc.** The addition of lead to Muntz metal improves considerably the machining properties, without significantly changing the strength and ductility. Leaded muntz metal has 60% copper, 0.3 to 0.8% lead and the remainder zinc.

(b) Tin bronzes : Copper - tin alloys are known as tin bronzes. Fig. 3.33 shows the thermal equilibrium diagram for such alloys. The dashed lines on the diagram indicate the phase that can occur with extremely slow cooling.

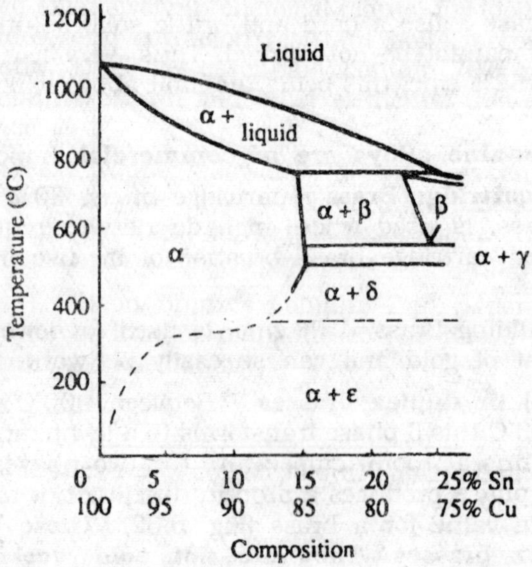

Fig. 3.33. Thermal equilibrium diagram for tin bronzes.

The structure that normally occurs with up to about 10% tin is predominantly α solid solution. Higher percentage tin alloys will invariably include a significant amount of the phase. This is a brittle intermeltallic compound, the phase being ductile.

(1) Phosphorus bronzes : Bronzes that contain up to 8% tin are bronzes and can be cold worked. In making bronzes, oxygen can react with the metals and lead to a weak alloy. **Phosphorus is normally added to the liquid metals to act as a deoxidizer. Some of the phosphorus remain in the final alloy. This type of alloy is known as phosphor bronze.** These alloys are used for springs, bellows, electrical contact, clips, instrument components. A typical phosphor bronze might have about 95% copper, 5% tin and 0.02 to 0.4% phosphorus.

The above discussion refers to wrought phosphor bronzes. Cast phosphor bronzes contain between 5 and 13% tin with as much as 0.5% phosphorus. A typical cast phosphor bronze used for the production of bearings and high grade gears has about 90% copper, 10% tin and a maximum of 0.5% phosphorus. This material is particularly useful for bearing surfaces; it has a low coefficient of friction and can withstand heavy loads. The hardness of the material occurs by virtue of the presence of both phase and a copper- phosphorus compound.

3.16 BEARING MATERIALS

The main requirements of a bearing materials are :

(1) Good compatibility : The material should have good compatibility which means good antiweld and antiscoring properties. Since most journal bearings will experience occasional metal to metal contact, **it is important that the bearing material be one that does not readily weld itself to the shaft material.**

(2) Conformability : Conformability is a measure of the bearings ability to adjust to misalignment or other geometric errors. Materials having a low modulus of elasticity usually have good conformability.

(3) Embeddability : Embeddability, means the material is soft enough to absorb foreign materials that are too large to pass through the oil film. By allowing this embedding, wear and scoring can be prevented.

(4) Good fatigure strength : The material should have good fatigue strength because of the many applications for which the bearing loads change directions.

(5) Good corrosion resistance : Resistance to corrosion is required in those applications where the lubricating oil used may oxidize and thus corrode the bearing.

(6) Wear resistance : The rate of wear can be an important factor in determining the life of a bearing.

(7) Strength and stiffness : The material must have a high elastic limit and strength to support the load, together with adequate modulus to resist deformation.

(8) Good shock resistance : The material must be tough and ductile to be able to withstand shocks.

(9) High thermal conductivity : The ability of the material to dissipate heat resulting from friction can be important in preventing a rise in temperature to a level which could result in melting of the bearing material and seizure of the bearing or degradation of a lubricant.

(10) Cost : Last, but certainly not least important, the cost should be as low as possible.

3.16.1 Types of materials

Let us now consider some of the more important types of materials.

(1) Babbitt alloys : Babbitt alloys, of either tin or lead base, are perhaps the **most commonly used material. They have excellent embeddability and compatibility characteristics.** However, babbitt alloys have the disadvantage of being relatively weaker than other bearing materials, lose strength rapidly with increasing temperature, and have lower fatigue strength. These shortcomings can generally be remedied by using a thin internal babbitt surface on steel backing.

(2) Copper alloys : Copper alloys are also widely used and are primarily bronzes. They are generally stronger, have a greater load capacity, have good fatigue strength, are harder, but less score resistant than babbitt bearings.

(3) Aluminium alloys : Because of their fatigue strength, load carrying capacity, thermal conductivity, corrosion resistance and low cost, are widely used for bearings in IC engines. Their disadvantages are that they require hardened journals and have poor compatibility, conformability, and embeddability. When a thin layer of babbitt is placed inside an aluminium bearing, the antiscoring and embeddability characteristics are improved.

(4) Porous metal: Porous metal bearing of either bronze or iron have found wide acceptance. These sintered - metal, self lubricated bearings have interconnecting pores in which oil is stored and then fed to the bearing surface. The low cost and long time between lubricant supply are their prime-advantages.

(5) Silver : Silver bearings are extremely useful for heavy duty applications. The bearings are made by electrodepositing silver on a steel backing with an overlay of lead 0.0254-0.127 mm (0.001-0.005 inch) thick. A thin layer of indium is then deposited on the lead to provide corrosion protection.

(6) Cast iron and steel : Cast iron and steel are used in bearings subjected to relatively light loads. The prime advantage of these bearings is low cost. However, because of poor embeddability and conformability, these bearings require very good alignment and freedom from foreign matter.

(7) Plastics : Various plastics such as phenolics, nylon, fluorocarbons, acetal and so on, are used as bearing materials. The advantages of using plastics are no corrosion, quiet operation, moldability, and **excellent compatibility. The last named characteristics usually means that no lubrication is required.** However, when used in heavy load application, their low heat conductivity may result in buildup of heat which will then require external cooling.

(8) Rubber : Rubber and other elastomeric materials are used in such applications as slip propellor shafts and rubber, hydraulic pumps, sand dredges, and so on. They provide vibration isolation, can compensate for misalignment and have good conformability. They are generally made fluted and placed inside a noncorrodible metal shell. A water flush will thus permit sand and grit to pass through the bearing without scoring it.

(9) Other materials : Other materials which are used in specially applications are wood (lignum vitae and oil-impregnated oak). ceramics and cermets, sapphire and glass.

EXERCISES

A. Theory

3.1 Why is the systematic classification of materials of importance to the materials scientist ?

3.2 Distinguish between elements, compounds and mixtures.

3.3 Distinguish between metals and non metals in terms of their characteristic properties and uses.

3.4 Briefly explain the meanings of the following terms as they relate to metals : alloy; clad metal, coated metal; non ferrous; sintered metal. Give examples to illustrate your answer.

3.5 Pure iron possesses three allotropic modifications. List these, together with their important physical, mechanical and magnetic properties.

3.6 Discuss the three main effects of increasing the percentage carbon in binary iron-carbon alloys from 0 to 6.67%.

3.7 Define the following terms : austenite, ferrite, cementite, pearlite, ledeburite.

3.8 Discuss the effect of carbon content on the structure and properties of steels.

3.9 Discuss the limitation of carbon steels on the engineering applications.

3.10 Sketch and label the steel section of the iron carbon system, using the term austenite, ferrite and cementite.

3.11 How do the structures of hypoeutectoid and hypereutectoid steels differ at room temperature as a result of being slowly cooled from the austenite state?

3.12 Explain how the percentage of carbon present in a carbon steel affects the mechanical properties of steel.

3.13 What is effect of tempering on martensite and hence the hardness of a material.

3.14 What factors limit the engineering applications of plain carbon steels?

3.15 In what ways do alloying elements, other than carbon, have an effect on the properties of steel.

3.16 Carbon steel is used for the following items. Which type of carbon steel would be most appropriate for each item?
(a) Railway track rails; (b) Ball bearings; (c) Hammers
(d) Reinforcement bars of concrete work; (e) Knives.

3.17 In what main ways do the following elements affect the properties of steel when they are alloyed with it ?
(a) manganese, (b) chromium, (c) molybdenum, (d) sulphur, (e) silicon.

3.18 What elements added to a steel can improve the corrosion properties of that steel?

3.19 What elements are usually added to a steel to increase its machinability?

3.20 Discuss the plain carbon steels classified by applications.

3.21 State two elements that, when added to steel, restrict grain growth.

3.22 What are stainless steels?

3.23 Explain how the three forms of stainless steel-ferritic, martensitic and austenitic are produced and how their properties differ.

3.24 Discuss the effects and functions of alloying elements.

3.25 What are the properties required of
(a) tool steels in general,
(b) tool steels for use in hot working processes,
(c) shock resistant tool steels?

3.26 Explain the effects on the microstructure and properties of cast iron of
(a) cooling rate,
(b) carbon content,
(c) the addition of silicon, manganese, sulphure and phosphorus.

3.27 Describe the conditions under which
(a) grey iron,
(b) white iron are produced.

3.28 In what way does the section thickness of a casting affect the structure of the cast iron?

3.29 How do the mechanical properties of malleable irons compare with those of grey irons?

3.30 Describe the way in which the structures of nodular irons and malleable iron differ from those of grey irons?

3.31 Which types of cast iron have high ductilities?

3.32 Which types of cast irons have high tensile strength?

3.33 Describe the forms of the microstructure of
(a) blackheart and
(b) whiteheart cast irons.

3.34 List the uses of various cast irons.

3.35 Why is it important to know whether a casting will require any machining before deciding on the material to be used ?

3.36 How does the presence of graphite as flakes or nodules affect the properties of the cast iron?

3.37 Which type of cast iron would be most suitable for a situation where there was a high amount of wear anticipated?

3.38 Which types of cast iron would you suggest for the following applications? Justify your answers.
(a) Sewage pipe
(b) Crankshaft in an IC engine.
(c) Brake disc
(d) Manhole cover.

3.39 Explain what is meant by the term critical points and, in this context, the symbols, A_1, A_3 and A_{cm}.

3.40 What are continuous cooling transformation curves?

3.41 Why must water quenching be used to harden plain carbon steels but some alloy steels can be hardened by air cooling?

3.42 Explain the term hardenability?

3.43 Explain the effect by
(a) Carbon
(b) Rate of cooling
(c) Different quenching media on the hardening of steel.

3.44 Describe the following treatment processes.
(a) Full annealing
(b) Process annealing
(c) Spheroidizing annealing
(d) Normalizing

3.45 Explain how flame hardening leads to the surface hardening of a steel.

3.46 What factors have to be considered in determining the type of quenching to be used to harden a steel?

3.47 What are the general effect of tempering martensitic steel?

3.48 Write the advantages of ferrous alloys over non ferous alloys.

3.49 Write the advantages of non ferrous alloys over ferrous alloys.

3.50 Describe wrought alloys which are not heat treatable.

3.51 Describe wrought alloys which are heat treated.

3.52 Describe three heat treatable wrought alloys (Duralumin, Magnesium - silicon alloy and copper - magnesium - zinc alloy) and their uses.

3.53 Describe Age hardening in aluminium alloys.

3.54 Describe precipitation hardening.

3.55 Describe non heat treated cast alloys.

3.56 Describe heat treated cast alloys.

3.57 Which form of copper should be used in an atmosphere containing hydrogen?

3.58 What is the effect of cold work on the properties of copper?

3.59 What is the effect of the percentage of zinc in a copper- zinc alloy on its strength and ductility?

3.60 What brass composition would be most suitable for applications requiring
(a) maximum tensile strength,
(b) maximum ductility,
(c) the best combined tensile strength and ductility.

3.61 The silver coinage is made from cupronickel alloy. What are the properties required of this material for such a use?

3.62 What are the constituent elements in
(a) brasses, (b) bronzes, (c) phosphorus bronzes, (d) aluminium bronzes, (e) cupronickels?

3.63 What are the constituent elements in
(a) cartridge brass,
(b) gliding brass,
(c) duplex brass,
(d) Muntz metal.

3.64 What is the reason that small percentage of lead is added in the Muntz metal.

3.65 What are the general characteristics of the nickel copper alloys are known as Monel?

3.66 What percentage of aluminium would be likely to be present in an aluminium bronze that is to be cold worked?

3.67 Describe Beryllium bronzes and silicon bronzes.

3.68 Write the application of copper alloys.

3.69 The following group of questions in concerned with justifying the choice of a particular alloy for a specific application.
(a) Why are milk bottle caps made of an aluminium alloy?
(b) Why are domestic water pipes made of copper?
(c) Why is cartridge brass used for the production of cartridge cases?
(d) Why are the ribs of hand gliders made of aluminium alloy?
(e) Why is brass used for cylinder lock keys?
(f) Why are kitchen pans made of aluminium or copper alloys?
(g) Why is copper used for gaskets?
(h) Why are electrical cables made of copper?
(i) Why are cupronickels used for tubes in desalination plants?

3.70 Compare the properties of ferrous and non ferrous alloys, commenting on the relative ease of processing.

3.71 List the main requirements of a bearing materials.

3.72 Classify the bearing materials.

3.73 Discuss the babbitt alloys used as bearing material.

3.74 Discuss the copper alloys used as bearing material.

3.75 Discuss the aluminium alloys used as bearing material.

3.76 Discuss the porous metal used as bearing material.

3.77 Discuss the silver as bearing material.

3.78 Discuss plastics as bearing material.

3.79 Discuss rubber as bearing material.

4

MANUFACTURING PROCESSES

4.1 INTRODUCTION

Deformation processes in manufacturing are operations that induce shape changes on the workpiece by plastic deformation under forces applied, by various tools and dies. Deformation processes can be grouped by temperature, size, and shape of the workpiece and type of operation. For example, using temperature as a criterion, deformation processes may be divided into three basic categories cold working (room temperature), warm and hot working.

Forging denotes a family of processes by which plastic deformation of the workpiece is carried out by compressive forces. Forging is one of the oldest metalworking operations known dating back 5000 B.C., and is used in making parts of widely varying sizes and shapes from a variety of metals. Typical parts made by forging today are crank shafts and connecting rods for engines, turbine disks, gears, wheels, bolt heads, hand tools, and many types of structural components for machinery and transportation equipment.

Sheet forming, involves, workpieces with a high ratio of surface area to thickness. A sheet thickener than 6mm is generally called a plate. Although relatively thick plates, such as those used in boilers, bridges, ships, and nuclear power plants, may have smaller ratios of surface area to thickness, sheet forming usually involves relatively thin material.

The products made by sheet forming processes include a large variety of shapes and sizes, ranging from simple bends to double curvatures with shallow or deep recesses. Typical examples are metal desks, appliance bodies, hubcaps, aircraft panels, beverage cans, car bodies and kitchen utensils. Thus sheet forming, which is also called pressworking, is among the most important of metal working processes. It dates backs to as early as 5000 B.C. When household utensils, jewellery, and other objects were made by hammering and stamping metals such as gold, silver, and copper.

Sheet metal is produced by a rolling process. If the sheet is thin, it is generally coiled; if thick, it is available as flat sheets or plates, which may have been decoiled and flattened.

Before a sheet metal part is formed, a blank of suitable dimensions is first removed from a large sheet. Removal is usually done by a shearing process. However, there are several other methods for cutting sheet, particularly plates. The sheet or plates may be cut with a band saw, which is a chip removal process. Flame cutting is another common method, particularly for thick steel plates, and is widely used in ship building and heavy structural components, Laser cutting has also become an important process and is used with computer controlled equipment to cut a variety of shapes consistently. In friction sawing, a disk or blade rubs against the sheet or plate at high surface speeds.

4.2 FORMING PROCESSES

The forming processes can be grouped under two broad categories, namely (i) hot forming and (ii) cold forming. **If the working temperature is higher than the recrystallization temperature of the material, then the process is called hot forming. Otherwise the process is termed as cold forming.** The flow stress behaviour of a material is entirely different above and below its recrystallization temperature. **During hot working a large amount of plastic deformation can be imparted without significant strain hardening.** This is important because a large amount of strain hardening renders the material brittle. **But in cold working plastic deformation of a metal takes place before the annealing temperature to cause permanent strain hardening.**

The frictional characteristics of the two forming process are also entirely different. For example, the **coefficient of friction in cold forming is generally of the order of 0.1, whereas that in hot forming can be as high as 0.6**. Further, hot forming lowers down the material strength so that a machine with a reasonable capacity can be used even for a product having large dimensions.

4.3 HOT WORKING

Hot working is the initial step in the mechanical working of most metals and alloys. **Hot working combines the working and annealing processes by deforming metal above the recrystallization temperature at which new grains are formed.** Since most metals and alloys have relatively high recrystallization temperatures, they must be worked at high temperatures.

4.3.1 Advantages and disadvantages of hot working

Advantages

(1) It can be used on most of metals, because it is a rapid and economical process.

(2) Due to refinements of grains, mechanical properties such as toughness, ductility, elongation and reduction in area are improved.

(3) Great latitude in shape and size of form is possible due to reduction of elastic limit.

(4) The power required to finish the part from ingot is less.

(5) The porosity of the steel ingot can be eliminated to greater extent.

(6) Directional property resulting from a fibre structure is obtained.

(7) A uniformity is established either by squeezing other impurities into fibre slags or distributed them throughout the mass.

(8). The grains are broken up and the parts are deformed into small and more numerous crystals or in other words the refinement of grain occur.

(9) The strength and hardness decreases at elevated temperatures.

Disadvantages

(1) The life of the tools used is less as they have to work at high temperature.

(2) Tooling and handling costs are high.

(3) There is a rapid oxidation or scaling of the surface due to high temperature of metal which results in poor surface finish.

(4) Close tolerances cannot be maintained; oxide films are surface characteristic.

4.4 HOT WORKING METHODS

The various methods of hot working are as follows:

(1) Rolling

(2) Forging

(3) Piercing or seamless tubing

(4) Drawing or cupping

(5) Spinning

(6) Extrusion

4.4.1 Rolling

Rolling is the most rapid method of forming metal into desired shapes by plastic deformation in between roll. The crystals are elongated in the direction of rolling, and they start to reform after leaving the zone of stress.

In deforming metal between rolls, the work is subjected to high compressive stresses as a result of the friction between the rolls and the metal. The friction forces are also responsible for drawing the metal into the rolls.

Traditionally, the initial material form for rolling is an ingot. However, this practice is now being rapid replaced by continuous casting and rolling, with much high efficiency and lower cost. Rolling is first carried out at elevated temperatures, wherein the coarse grained, brittle and porous structure of the ingot or continuously cast metal is broken down into a wrought structure, with finer grain size (Fig 4.1)

A wide variety of rolling equipment is available with a number of roll arrangements, the major types of which are shown in Fig. 4.2. Two-high or three-high (developed in the mid 1800s) rolling mills are used for initial breakdown passes on cast ingots with roll diameters ranging from 600 to 1400 mm. Small-diameter rolls are preferable,

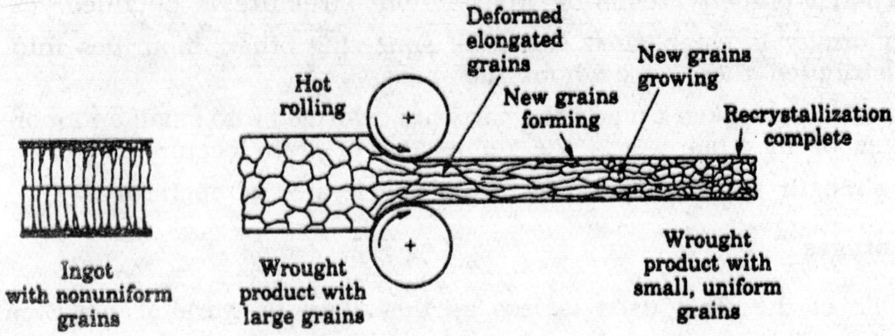

Fig. 4.1. Hot rolling

because the smaller the roll radius the lower the roll radius the lower the roll force. However, small rolls deflect under roll forces and have to be supported by other rolls [Figs. 4.2 (c) and (d)]

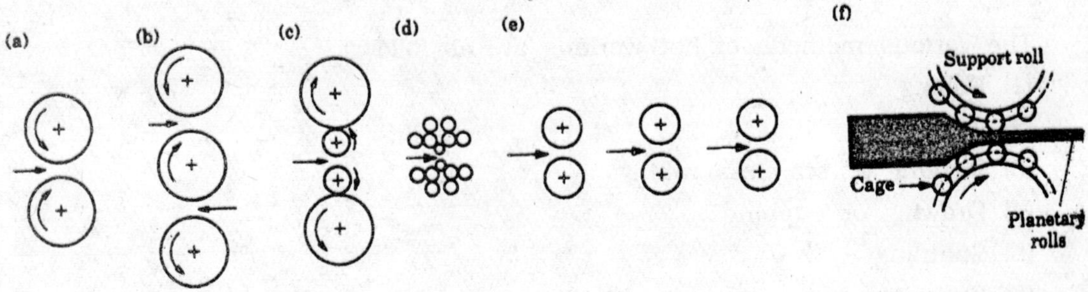

Fig. 4.2. Schematic illustration of various roll arrangements (a) 2 high (b) 3 high (c) 4 high (d) cluster.

The range of dimensions in rolling is extremely wide. The rolled product can be as wide as 5000 mm and as thin as 0.0025 mm obtained in a cluster mill.

Stiffness in rolling mills is important for controlling dimension. Mills can be highly automated with rolling speeds as high as 25m/s. Flat rolling can also be carried out with front tension only, using idling rolls (steckel rolling). For this case, the torque on the roll is zero, assuming frictionless bearings.

Requirements for roll materials are mainly strength and resistance to wear. Three common roll materials are cast iron, cast steel, and forged steel. For hot rolling, roll surfaces are generally rough and may even have notches or grooves in order to pull the metal through the roll gap at high reductions. Rolls for cold rolling are ground to a fine finish and, for special applications, are also polished.

4.4.2 Forging

Most of the forging produced by a black smith are the result of a sequence of fairly standard operations.

Forging can be defined as the controlled plastic deformation of metals at elevated temperatures into a predetermined size or shape using compressive forces exerted through some type of die by a hammer, a press or an upsetting machine.

Advantages of forging

(1) Forging improves the structure of the metal and hence its mechanical properties.

(2) Forging distorts the previously created uni-direction fibre in such a way so that it strengthen the component.

(3) The forging have high strength and ductility and offer great resistance to impact and fatigue loads due to extra working during the process and the opportunity of aliging the grain flow.

(4) Forging renders the parts uniform in density as well as dimensions.

(5) Forging allows the material to be displaced where it is needed, thus effecting a considerable weight and cost reduction.

Disadvantages of forging

(1) The initial cost of dies and the cost of their maintenance is high.

(2) Forging operation is limited to simple shapes and has limitation for parts having under-cuts, re-entrant surfaces etc.

(3) The rapid oxidation of metal surface at high temperature results in scaling which wears the dies. Some materials are not readily worked by forging.

Methods of forging

There are many variations of the basic forging operation, and the most commonly practised are (i) smith forging, (ii) drop forging, (iii) press forging (iv) upset forging, (v) swaging, and (vi) roll forging.

(i) In Smith forging, a hot workpiece is give the desired shape by using hand-held tools and hammers. Nowdays, power-driven hammers are used to impart the repeated blows. The anvil and the hammer are mostly flat and the desired shape of limited varieties is obtained by a manipulation of the job between the blows.

(ii) In drop forging, the impact loads (blows) are applied to the workpiece to cause metal flow for filling up the cavity formed by the two halves of the closed die. To ensure complete filling normally an excess amount a material is provided. This excess material flows out circumferential to form a flash which is subsequently trimmed. When the product geometry is complicated, a set of dies may be necessary to obtain the final form.

(iii) Instead of the repeated blows, a gradual force is applied in **press forging**. However, depending on the complexity of the job, a set of dies may be required to obtain the final product.

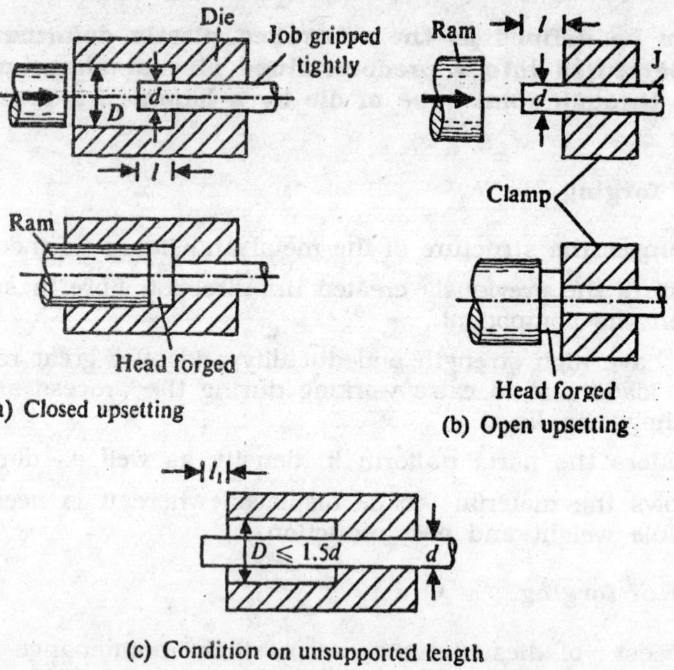

Fig. 4.3. Upsetting operation

(iv) A localized forging operation is commonly known as **upsetting.** The upsetting operation may be both closed and open, as shown in Figs. 4.3a and 4.3b, respectively. Clearly, the operation involves a longitudinal compression of the bar stock.

(v) Swaging is a special variation of impact forging where the repeated blows are obtained by a radial movement of shaped dies, as explained in Fig. 4.4. This operation is generally used for reducing the diameters and tapering of bars and tubes.

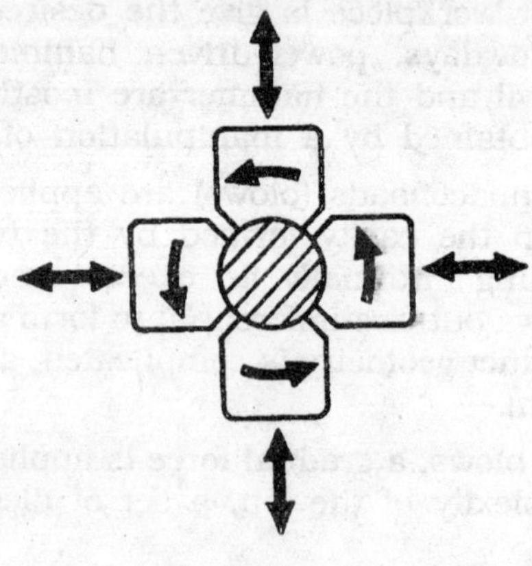

Fig. 4.4. Principle of rotary forging.

(vi) Roll forging is performed with two semicircular, grooved rolls held by two parallel shafts, as shown in Fig. 4.5. The process is used for reducing the diameter of rods.

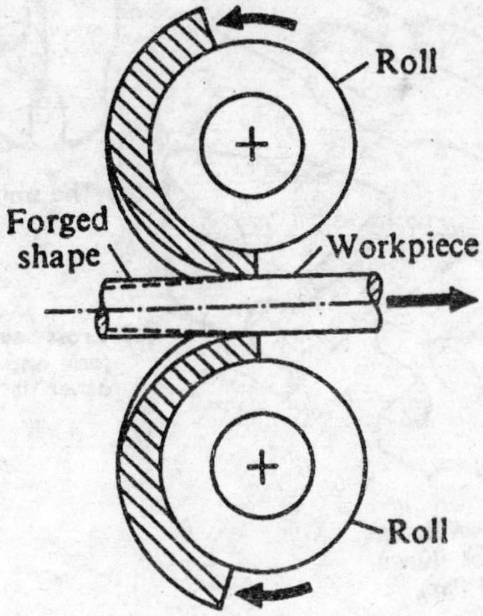

Fig. 4.5. Principle of roll forging.

The heated workpiece is placed between the dies in an open position. After a half revolution of the rolls, the workpiece is rolled out. It is then put in the smaller groove and the operation continued until the desired dimension is achieved.

4.5 FORGING TOOLS

The anvil (Fig. 4.6) is for supporting the work whilst it is being struck with the hammer, as well as providing means for other forging operations. The body of the anvil is of mild steel and, to give a hard top face, a piece of high carbon steel about 20 mm to 25 mm this is welded on. The beak is soft like the anvil body and its shape makes it useful for bending metal to rounded formations. The ledge between the beak and the anvil face is also soft and may be used for resting metal when cutting through with a chiesel; the soft underneath metal does not damage the chiesel

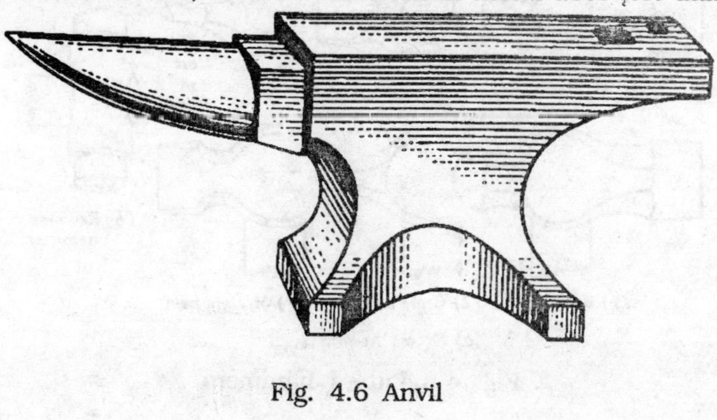

Fig. 4.6 Anvil

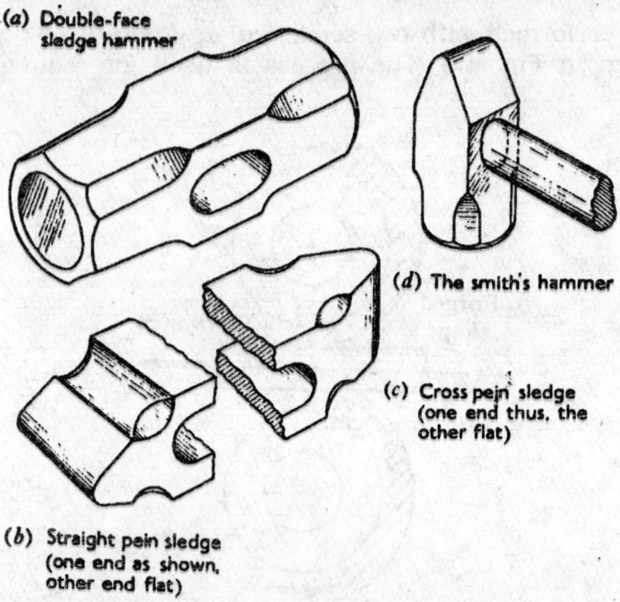

Fig. 4.7 Black smith's hammers

edge. In the top of the anvil is a square hole to the shank of various tools. Anvils vary up to about 150 kg and should stand with the top face about 600 mm from the fuller. This height may be attained by resting the anvil on a cast iron or wood base.

Two kinds of hammer are used in hand forging :

(1) the hand hammer used by the black smith himself;

(2) the sledge hammer used by the striker; the black smith's assistant.

Hammers particular to the black smith are shown in Fig. 4.7, whilst engineer's hammers, which he also uses, are in Fig. 4.8.

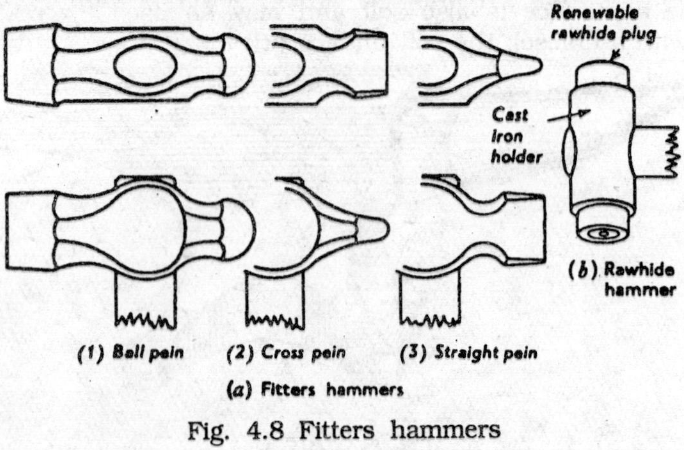

Fig. 4.8 Fitters hammers

Hammers heads should be cast steel, the ends hardened and tempered, with the portion round the eye soft. Smith's hand hammers usually have a slightly convey striking face and should be from 1 kg to 1.5 kg in mass. Sledge hammers vary from 4.5 kg to 5.5 kg for ordinary work to 7 kg to 9 kg for heavy blows, the shaft being about 1 m long. When the smith and his strikers are at work, the smith indicates with his hand hammer the point where he requires the sledge to fall, and when he allows his hammer to ring on the anvil it is an indication that the striker is to cease for the time being.

4.6 HAND TOOLS

For roughly forming the metal, direct blows from the hammer alone may be used, but for cutting off, forming and other finishing operations, various hand tools are necessary. The most common of these are as follows :

4.6.1 Chisels

These are used for cutting metals and for nicking prior to breaking. They may be hot, or cold, depending on whether the metal to be cut is hot or cold, and the main difference between the two is in the edge. The cold chisel has its edge hardened and tempered with an angle of about 60°, whilst the angle of the hot chisel is 30° and hardening is not necessary. Since in any case the hot metal would re-soften it. The edge of the chisel should not be quite straight but slightly rounded as shown in Fig. 4.9.

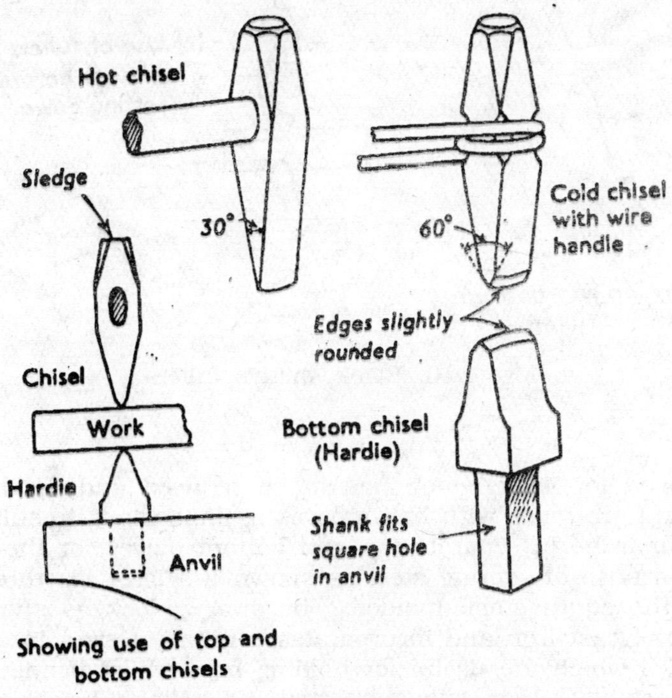

Fig. 4.9. Black smith's chisel.

Chisels are generally used in pairs, a pair comprising a top tool and bottom tool (often called the hardie). The hardie has a square shank and fits in the square hardie hole in the anvil face, whilst the top chisel, which is held by the smith and hit by the striker, may be fitted with either a wooden handle or a metal wire handle as shown in Fig. 4.9.

4.6.2 Fullers

Fullers are used for necking down a piece of work, the reduction often serving as the starting point for a reduction as shown in Fig. 4.10. They are made in top and bottom tools as in the case of chisels, the bottom tool fitting in the hardie hole and the top held by the smith and struck by the striker. Fullers are made in various sizes according to needs, the size denoting the width of the fuller edge. Thus a 12 mm fuller would have a semi-circular edge 12 mm wide. For shouldering round work fullers may have their edges hollowed out as shown in Fig. 4.10(b).

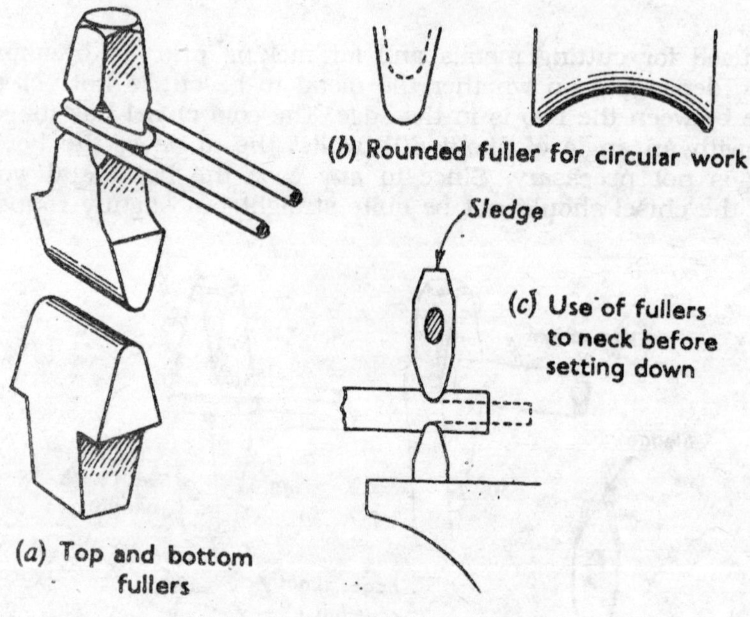

Fig. 4.10. Black smith's fullers

4.6.3 Swages

Swages are used for work which has to be reduced and finished to round or hexagonal form and are made with half grooves of dimensions to suit the work being reduced. Swages may be in separate top and bottom halves, or the two halves may be connected by a strip of spring steel as shown in Fig. 4.11, thus enabling them to be used by the smith single-handed. The swage block as shown in Fig. 4.12 is generally made of cast iron and incorporates a range of sizes in addition to being provided with holes which are useful for holding bars whilst bending, and knocking up heads. The swage block is usually supported at a suitable height on a stand which is adaptable to hold its flatwise, or on its edge.

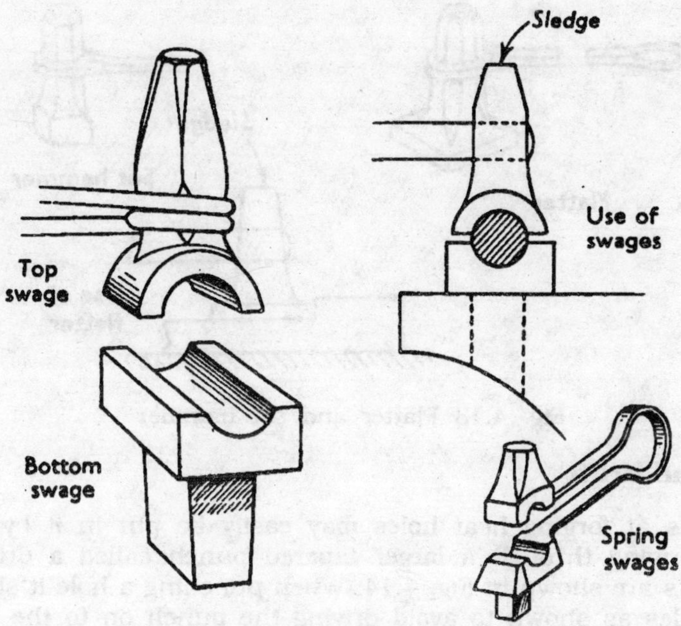

Fig. 4.11 Black smiths swages

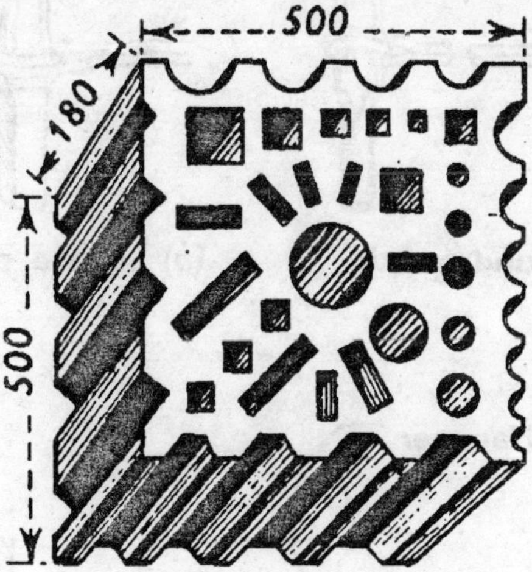

Fig. 4.12 Swage block

4.6.4 Flatters

Flatters are used for finishing flat surfaces and are made with a perfectly flat face about 75 mm squares (or round) as shown in Fig. 4.13. The set-hammer is a similar but smaller tool used for finishing in corners and confined spaces. As the work is supported directly on the anvil for flatting, only the top, handled tool is necessary.

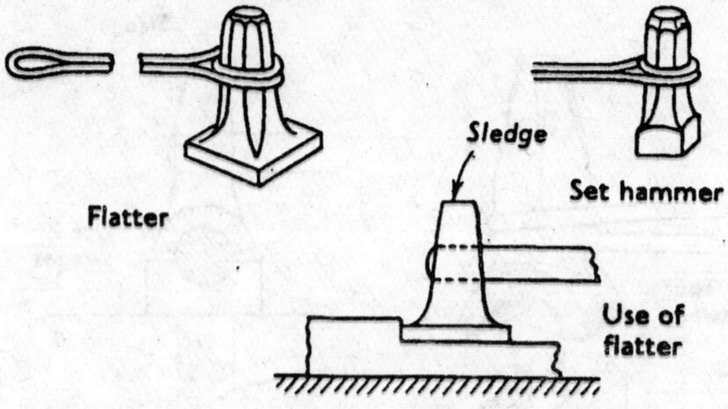

Fig. 4.13 Flatter and set hammer

4.6.5 Punches and drifts

When metal is at forging heat holes may easily be put in it by punching, and opened out by driving through a larger tapered punch called a drift. Examples of punches and drifts are shown in Fig. 4.14. When punching a hole it should be carried out from both sides as shown to avoid driving the punch on to the hard anvil face.

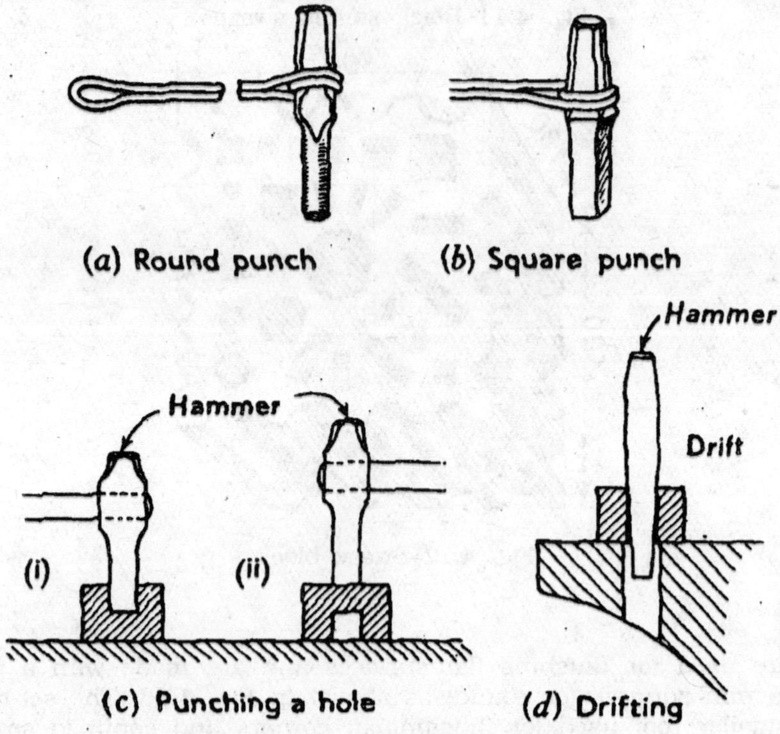

Fig. 4.14 Punches and drifting.

4.6.6 Tongs

The smith requires a good selection of tongs with which to hold this work during the various operations. The chief types of tongs for holding work are shown in Figs. 4.15 (a), (b) and (c).

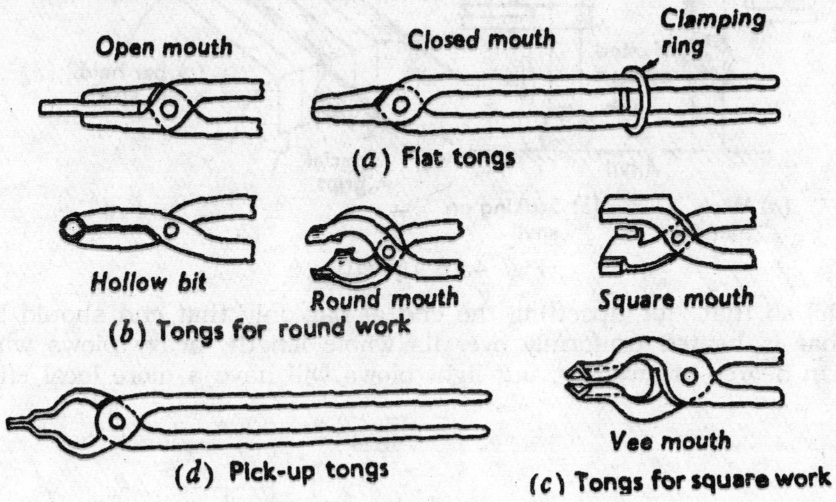

Fig. 4.15 Black smith's tongs

Those in Fig. 4.15 (a) are the flat tongs and should hold along the entire length of the jaws. If they grip at the front or back only the grip will not be secure. Hollow tongs for round work are shown in Fig. 4.15 (b) and tongs for square bars in Fig. 4.15 (c). In these types there should, of course, be a variety of sizes suitable for dealing with the range of work to be undertaken. The tongs shown in Fig. 4.15 (d) are for picking up round bars, but not for holding work during forging. To relieve the smith of the strain of holding the tongs during a long forging operation, means are often provided for clamping the tongs together, and one method is to drive a loop over the handles as shown in Fig. 4.15.

4.7 FORGING OPERATIONS

The formation of a shape by forging consists of a combination of two or more of a number of relatively simple operations which are as follows :

4.7.1 Upsetting

This consists of increasing the thickness of a bar at the expense of its length and is brought about by end pressure. The pressure may be obtained by driving the end of the bar against the anvil, by supporting on the anvil and hitting with the hammer, by placing in swage block hole and hitting with hammer or by clamping in a vise and hammerling as shown in Fig. 4.16. The position and nature of the upsetting will depend on the heating, and upon the type of blow delivered. In general, the increase in lateral swelling will be greatest at the parts where the metal is hottest

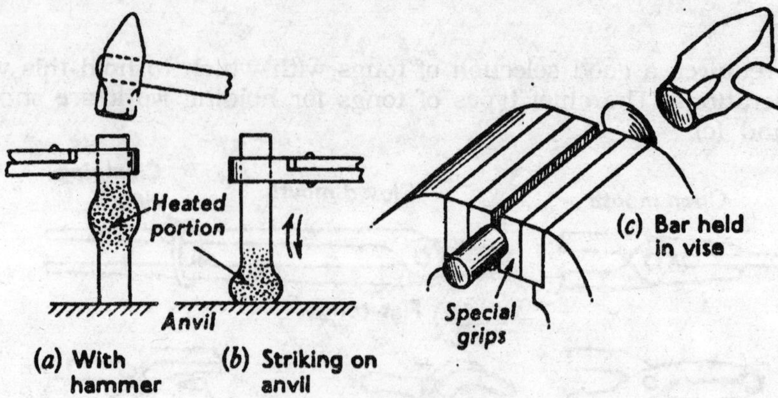

Fig. 4.16 Upsetting

(more plastic) so that for upsetting the end of bar only that end should be heated. If a short bar is heated uniformly over its whole length, heavy blows will cause a fairly uniform degree of swelling, but light blows will have a more local effect at the ends only.

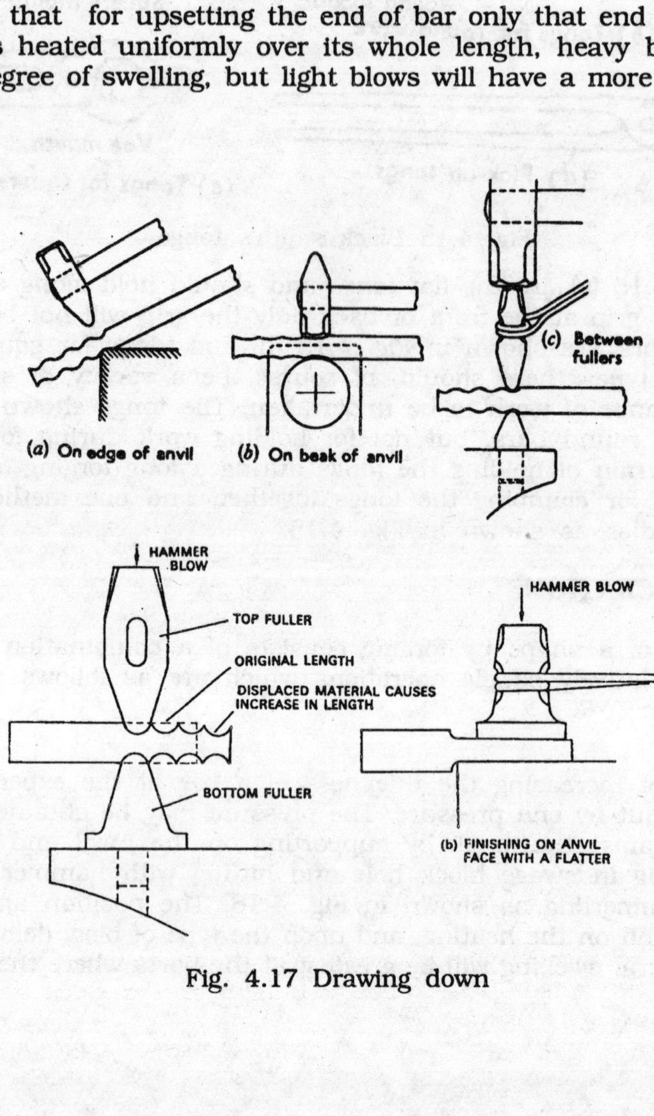

Fig. 4.17 Drawing down

4.7.2 Drawing down

This is the opposite of upsetting i.e. the section of the bar is reduced locally, with increase in length at the expense of its width or thickness or both. The point where drawing down is to start is usually first notched, necked by fullers, and drawing down is continued from that point either between a pair of fullers or against the corner edge of the anvil. This leaves an undulating surface which is usually finished with a flatter or set hammer, as shown in Fig. 4.17.

A good method of drawing down square or rectangular bars is to use the edge or beak of the anvil as shown in Fig. 4.17 (a) and (b), turning the bar through 90° if the thickness in both directions is to be reduced. When assisted by a striker, a pair of fuller may be used as shown in Fig. 4.17. When the preliminary drawing has been done in this way the work may be finished off with the flatter.

When round bars have to be drawn down a considerable amount they should first be brought down to square, drawn by the above method and then squared up again. The square is then taken to an octagon by taking off the corners and finally rounded and finished between stages. When reducting the thickness of metal care should be taken not to drive it outwards from the centre too rapidly or internal cracks will occur which will remain as permanent weakness in the metal.

4.7.3 Setting down

Setting down is a local thinning down effected by the set-hammer or set. Usually the work is fullered at the place where the setting down commences as shown in Fig. 4.18.

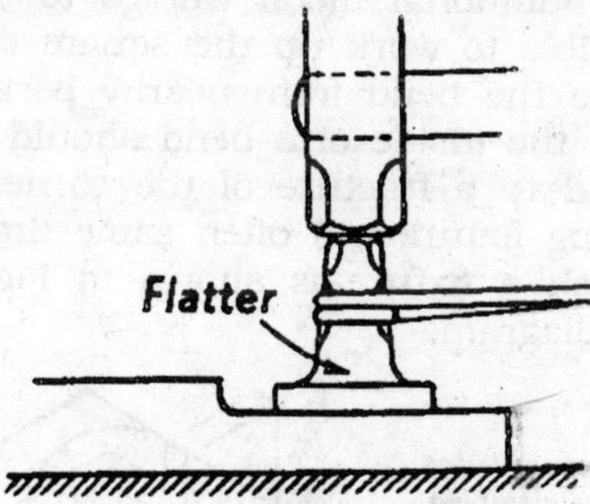

Fig. 4.18 Setting down

4.7.4 Bending

Bending is an important operation in forging and is one very frequently used. Bends may be either sharp-cornered angle bends or they may be composed of a more gradual curve. Angle bends may be made by hammering the metal over the edge of the anvil, over a block of metal held in the hardie hole or in a vise, or over a vise jaw itself, whilst the metal is being gripped as shown in Fig. 4.19 (a).

When metal is bent the layers of metal on the inside are shortened and those on the outside are stretched. This causes a bulging of the sides at the inside, and

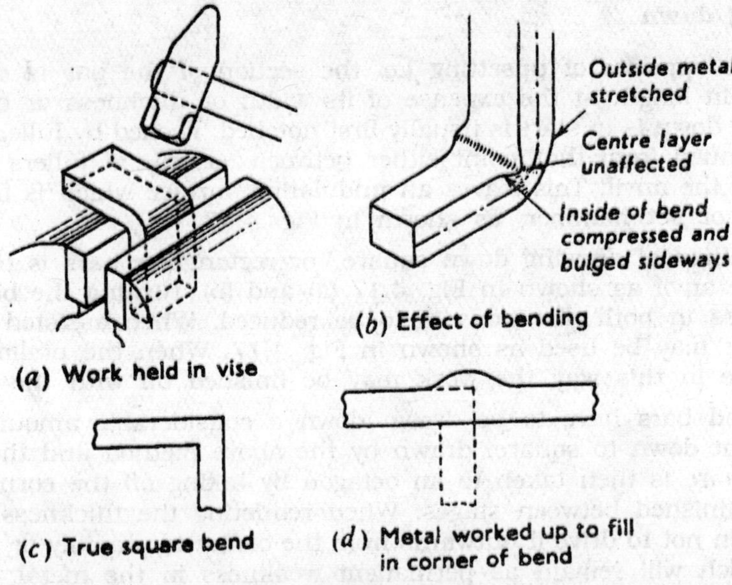

Fig. 4.19 Bending (Sharp bends)

a radius on the outside of the bend as shown in Fig. 4.19 (b). If a true square bend is required as shown in Fig. 4.19 (c), additional metal must be worked to the place where the bend occurs as shown in Fig. 4.19 (d).

When this is bent the additional metal will go to make up the corner. If this is not done it may be possible to work up the square corner by blows from a light hammer, working metal into the bend from nearby portions of the work. Whenever possible, a sharp corner on the inside of a bend should be avoided as it constitutes a weakness which may lead to a fracture of the corner. When double bends have to be made a simple bending fixture will often same time and enable more uniform results to be obtained. Such a fixture is shown in Fig. 4.20 and its operation is self explanatory from the diagram.

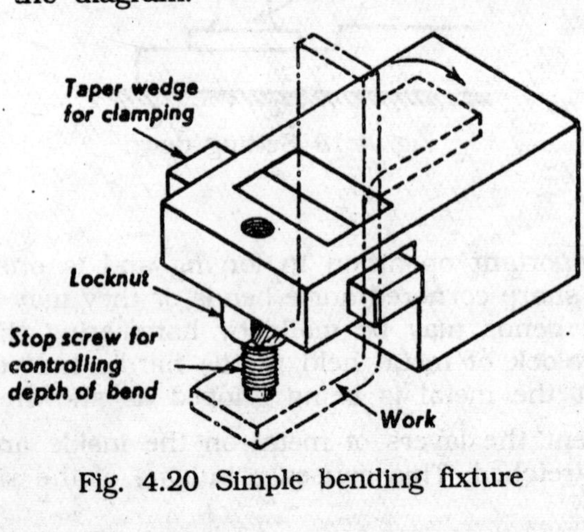

Fig. 4.20 Simple bending fixture

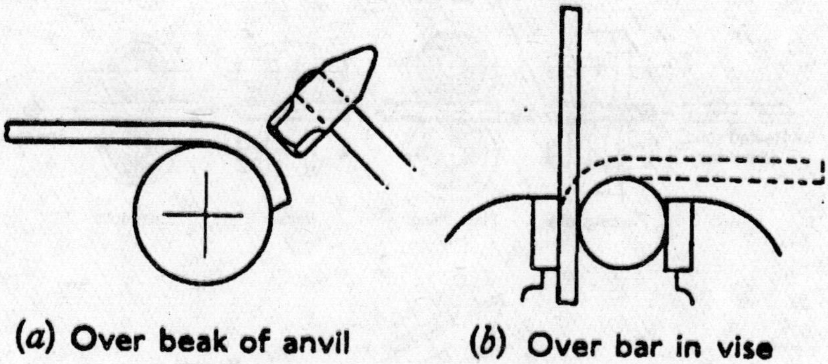

Fig. 4.21 Circular bending

Gradual bends may be made by using the beak of the anvil as a former as shown in Fig. 4.21 (a), or the metal may be bent round a bar of the correct radius held in a vise as in Fig. 4.21 (b), When a quality of bending has to be done, the use of some type of bending fixture will same time and produce better and uniform bends. In Fig. 4.22 the handle carrying the roller is swung round, causing the metal to be bent round the centre disc. Varying sizes of bends may be obtained by fitting different centre pieces and rollers to suit.

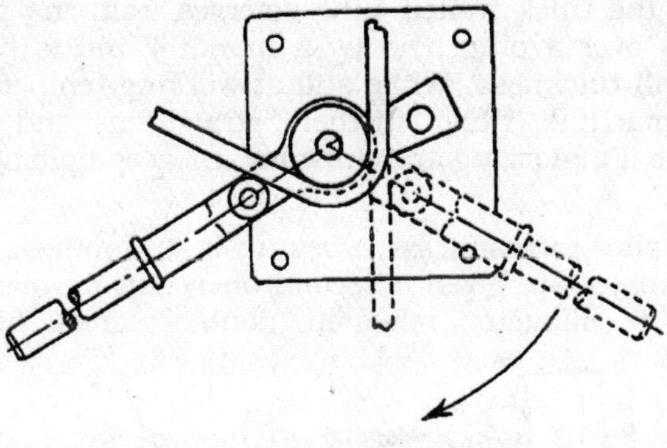

Fig. 4.22 Bending fixture.

4.7.5 Piercing

This is the most economical mechanical working process for the manufacture of seamless tubes. In this process hot billet (950°C to 1100°C) is forced over the mandrel with the help of two conical shaped rollers as shown in Fig. 4.23 (a). These conical shaped rolls operating in the same direction. Because these rolls is a fixed point or mandrel that assists in the piercing and controls the size of the hole as the billet is forced over it.

The entire operation of making seamless tubing by this process is shown in Fig. 4.23. The solid billet is first centre punched and then brought to forging heat in a furnace before being pierced. It is then pushed into the two piercing rolls, which impart both rotation and axial advance. The alternate squeezing and bulging of the

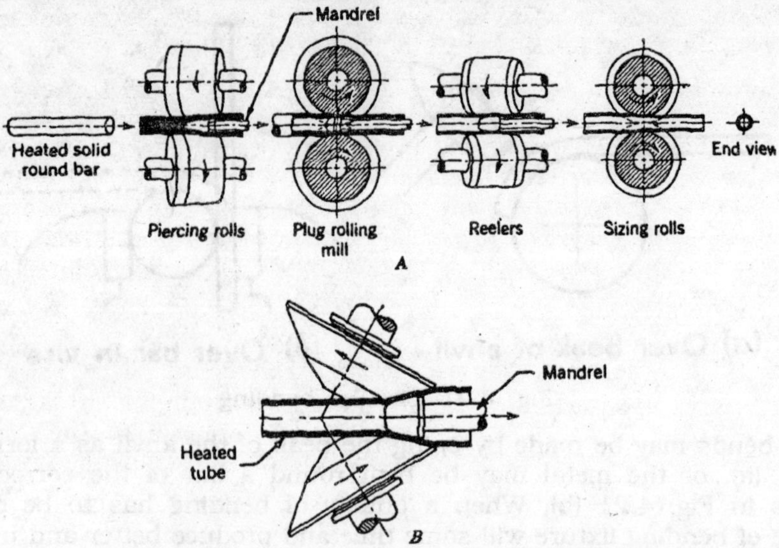

Fig. 4.23 (a) Principal steps in the manufacture of seamless tubing. (b) Rotary seamless process for large tubing.

billet open a seam in its centre, the size and shape of which are controlled by the piercing mandrel. As the thick walled tube emerges from the piercing mill, it passes between grooved rolls over a plug held by a mandrel and is converted into a longer tube with specified wall thickness. While still at working temperature, the tube passes through the reeling machine, which further straightens and sizes it and gives the walls a smooth surface. Final sizing and finishing are accomplished in the same manner as with welded pipe.

This procedure applies to seamless tubes up to 150 mm in diameter. Larger tubes up to 355 mm in diameter are given a second operation on piercing rolls. To produce sizes up to 610 mm in diameter, reheated, double pierced tubes are processed on

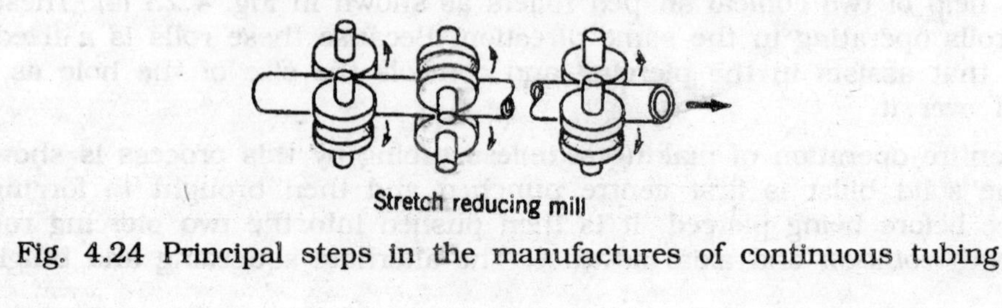

Fig. 4.24 Principal steps in the manufactures of continuous tubing

a rotary rolling mill, as shown in Fig. 4.23 (b) and are finally completed by reelers and sizing rools, as described in the single piercing process.

In the continuous method shown in Fig. 4.24, a 139.7 mm round bar is pierced and conveyed to mandrel mill, where a cylindrical bar or mandrel is inserted. These rolls reduce the tube diameter and wall thickness. The mandrel is then removed and the tube reheated before it enters a stretch-reducing mill. This mill reduces not only the wall thickness of the hot tube also the tube diameter. Each successive roll speeds up to produce a tension sufficient to stretch the tube between stands. The maximum delivery of this mill is 6.6 m/s for a pipe around 50 mm in diameter.

4.7.6 Drawing

Drawing is defined as a process of making cup-shaped parts from sheet metal blanks by pulling it into dies with the help of punch. Sheet-blanks are heated to working temperature to provide sufficient plasticity. These are then placed on the die or cavity as shown in Fig. 4.25 and the punch is hammered over the heated blank.

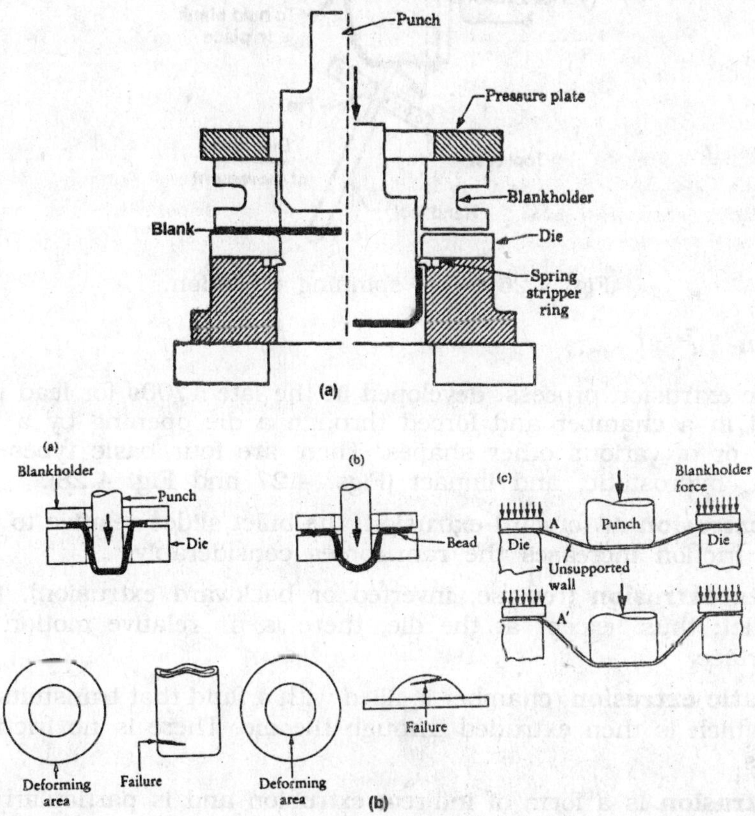

Fig. 4.25. Principle of drawing (deep drawing) Examples of drawing operations (a) Pure drawing (b) Pure stretching (c) Possibility of wrinkling in the unsupported region of a sheet in drawing

Deep drawing, first developed in the 1700s, has been studied extensively and has become an important metalworking process. Typical parts produced are beverage can, pots and pans, containers of all shapes and sizes, sinks and automobile panels.

4.7.7 Spinning

Metal spinning is the operation of shaping thin metal by pressing it against a form while it is rotating. The nature of the process limits it to symmetrical articles. This work is done on a speed lathe, similar to the wood lathe, except that in place of the usual tailstock, it is provided with same means of holding the work against the form, as shown in Fig. 4.26. The forms are usually turned from hardwood and attached to the face plate of the lathe, although smooth steel chucks are recommended for production job.

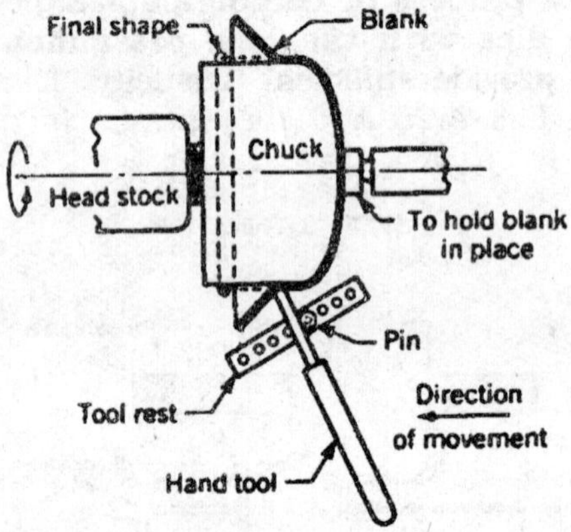

Fig. 4.26 Metal spinning operation.

4.7.8 Extrusion

In the basic extrusion process, developed in the late 1700s for lead pipe, a round billet is placed in a chamber and forced through a die opening by a ram. The die may be round or of various other shapes. There are four basic types of extrusion: direct, indirect, hydrostatic, and impact (Figs. 4.27 and Fig. 4.28).

In **Direct extrusion** or forward extrusion, the billet slides relative to the container wall, the wall friction increases the ram forces considerably.

In **Indirects extrusion** (reverse, inverted or backward extrusion), the die move toward the billet; thus, except at the die, there is no relative motion at the billet container interface.

In **hydrostatic extrusion**, chamber is filled with a fluid that transmits the pressure to the billet, which is then extruded through the die. There is no friction along the container walls.

Impact extrusion is a form of indirect extrusion and is particularly suitable for hollow shapes.

An interesting example of impact extrusion is in the manufacture of collapsible tubes for shaving cream, toothpaste, and paint pigments. These extremely thin tubes

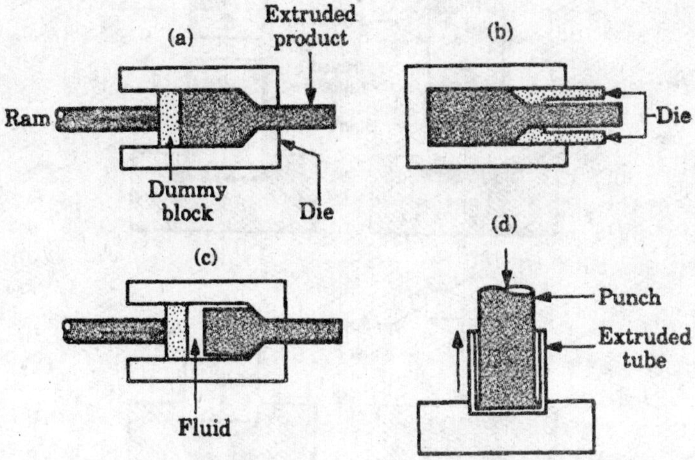

Fig. 4.27 Types of extrusion (a) direct, (b) indirect; (c) hydro static (d) impact.

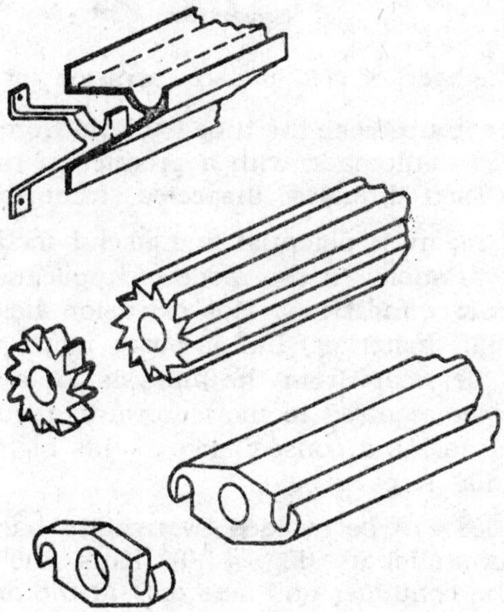

Fig. 4.28. Extrusions and examples of product made by section of extrusions.

are pressed from slugs, as illustrated in Fig. 4.29. The punch strikes a single blow, causing the metal to squirt around the punch.

The outside diameter of the tube is the same as the diameter of the die, and the thickness is controlled by the clearance between the punch and die. The tube shown in Fig. 4.29 has a flat end, but any desired shape can be made by properly forming the die cavity and the end of the punch. For toothpaste tubes, a small hole is punched in the centre of the blank and the die cavity is shaped to form

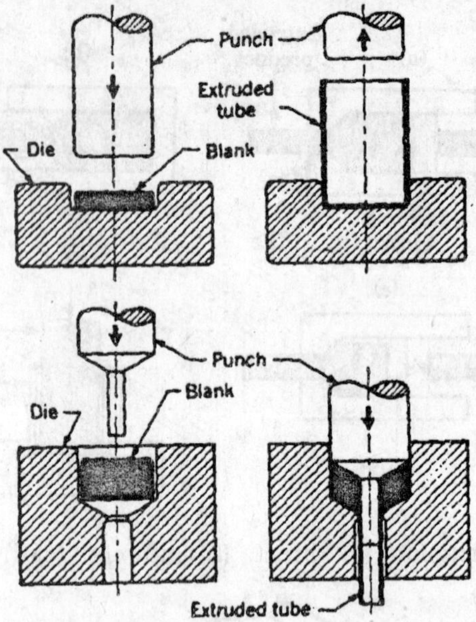

Fig. 4.29 Methods of cold impact extrusion for soft metals.

the neck of the tube. On the upstroke the tube is blown from the ram with compressed air. The entire operation is automatic with a production rate of 35 to 40 tubes per minutes. The tubes are then threaded, inspected, trimmed, enameled, and printed.

Extrusion is one of the most potential and useful metal working processes and has a large number of variations in the mode of application. **It can be performed under both hot and cold conditions.** Hot extrusion helps reduce the work load (specially for high strength materials) but it poses more problems such as cooling arrangement and rapid die wear; from the analysis for a simple forward extrusion process, the whole billet is required to move forward, resulting in a large frictional loss and high working load. As a consequence of this high work load, the container is subjected to high radial stresses.

The forgoing difficulties can be avoided by using a backward extrusion process, where the billet remains stationary (Fig. 4.30). Thus, the frictional force is absent between the billet and the container and acts only at the die-container interface. The

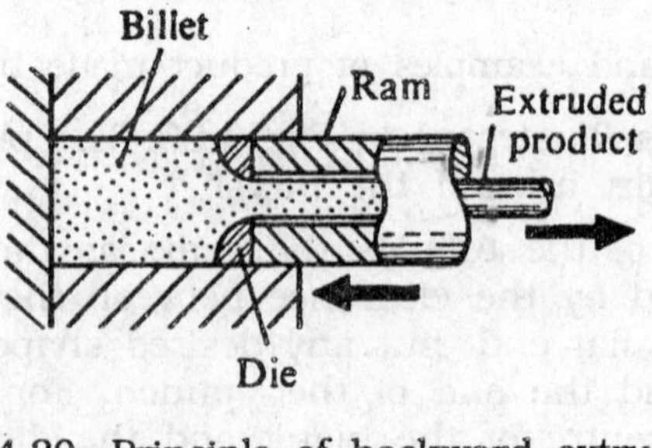

Fig. 4.30. Principle of backward extrusion.

magnitude of the latter is much less than that of the frictional force encountered in a forward extrusion process. Hence, the work load is reduced and also it is independent of the billet length.

Tubular sections can also be extruded by using a mandrel along with the ram, as illustrated in Fig. 4.31. Both open (Fig. 4.31 a) and closed (Fig. 4.31 b) end products can be obtained depending on the initial blank shape. The mandrel may, either be fixed to the ram or to a separate body, as indicated in Fig 4.31c.

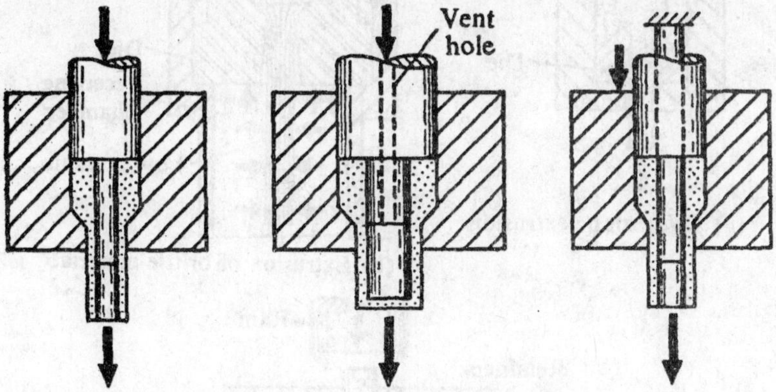

Fig. 4.31. Methods of extruding tubular section.

Thin-walled cans may be obtained by using impact extrusion. This process is limited to soft and ductile materials and is normally performed under cold conditions. The process is schematically shown in Fig. 4.32.

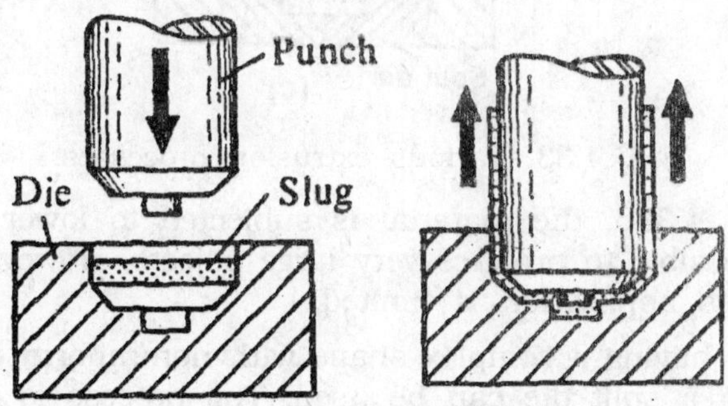

Fig. 4.32. Impact extrusion process (cold working)

Instead of applying the load on the billet directly by the ram, a fluid medium can be used, as illustrated in Fig 4.33a. This process is known as hydrostatic extrusion; here, the frictional loss at the billet container interface is, eliminated. A slight variation of this process offers a possibility of extruding a relatively brittle material. In this, apart from the large hydrostatic pressure applied to the billet, the product in the receiving chamber is maintained under a lower pressure (about one-half the pressure applied to the billet).

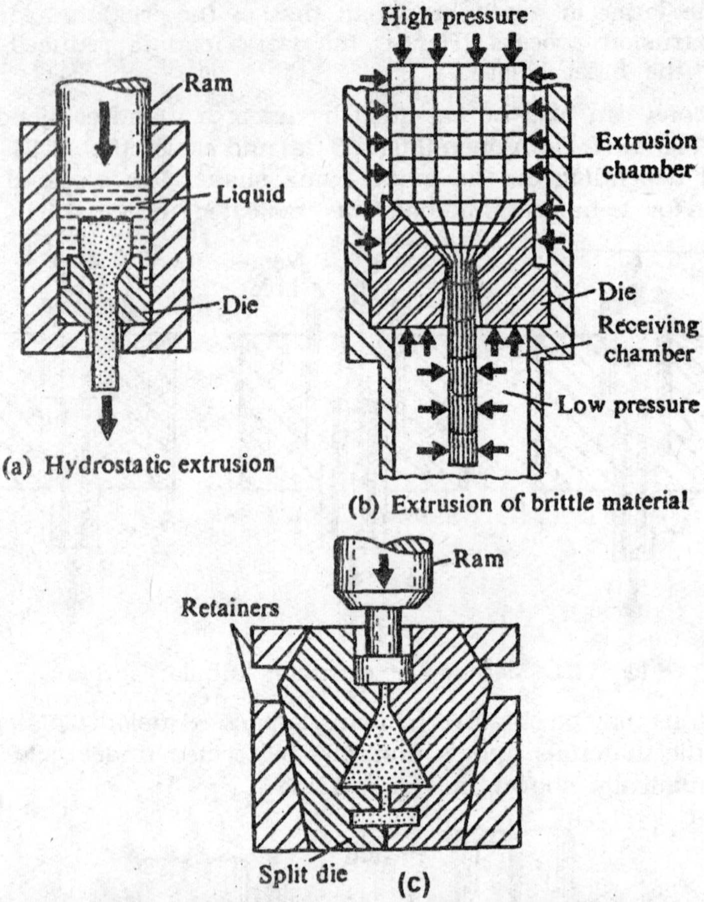

Fig. 4.33. Various extrusion processes.

As shown in Fig 4.33b, the material is subjected to lower strain gradients. In this process, it is possible to produce very large objects, However, since the process is inherently slow, its application is limited.

To produce a job having a complex shape with nonuniform cross- section, closed cavity extrusion with a split die can be used. The process is similar to closed die forging and is illustrated in Fig 4.33c.

The temperature range of the billet during the hot extrusion of steel is 1200-1500°C. The die must be kept at a lower temperature (approximately 200°C) to avoid excessive wear rate. Glass fibres (or powders) are normally used as lubricants since the viscosity of glass is sensitive to temperature. Thus, the viscosity is high at the die surface, providing a good protection to die wear and faciliating the formation of a glass skin (about 0.025 mm thick) on the product. At the same time, the work load is reduced since the viscosity of glass is much lower at the billet - container interface.

4.8 COLD WORKING

The working of metals at temperatures below their recrystallization temperature is defined as cold working. Most of the cold working processes are performed at room temperatures. Even though the forming does cause a temperature rise, it is not as much as hot working. Hot working on the other hand, is performed on metal in the plastic state and refines the grain structure. **Cold work distorts the grain and does little toward reducing the size of the material. Cold work improves the strength, machine ability dimensional accuracy and surface finish of the metal.**

4.8.1 Principles of cold working

When material is cold worked, the change in material shape brings about marked changes in the grain structure. Structural changes that occur are grain fragmentation movement of atoms, and lattice distortion.

Much greater pressures are needed for cold working than for hot working. The metal being in a more rigid state, is not permanently deformed until stresses exceeding the elastic limit are passed. Because there can be no recrystallization of grains in the cold working range, there is no recovery from grain distortion or fragmentation. As grain deformation proceeds, greater resistance to this action results in increased strength and hardness of the metal. The metal is said to be strain hardned and for some metals that will not respond to heat treating. In general, they refer to resistance built up in the grains by atomic dislocation, fragmentation or lattice distortion or a combination of all these phenomena.

The amount of cold work that a metal will stand is dependent on its ductility. The higher the ductility of a metal the more it is able to be cold worked. Pure metals withstand a greater amount of deformation than metals having alloying elements because alloying increases the tendency and rapidity of hardening. Large grain metals are more ductile than small ones. When metal is deformed by cold work severe stresses known as residual stresses are set up in the metal. To remove these undesirable stresses the metal must be reheated to slightly below the recrystalline range temperature. In this range the stresses are rendered ineffective without appreciable change in physical properties or grain structure Heating into the recrystallization or annealing range eliminates the effect of cold working. Sometimes it is desirable to have residual stresses in the metal. The fatigue life of small parts may be improved by shot peening, which causes the surface metal to be in compression and the material below the surface to be in tension.

The cold worked condition is described as the annealed (soft) temper, quarter-hard, half-hard, three-quarter hard, full hard and spring temper. Each temper condition indicates a different percentage of cold reduction following the annealing treatment. The increase in hardness resulting from cold working is described as work hardening.

4.8.2 Advantages and Limitations of cold working

The advantages and limitations of the cold working on metals are given below:

Advantages

(1) Surface finish is improved and close tolerance can be maintained.

(2) Distortion or fragmentations of the grain structure is created.

(3) Accurate dimensions of parts can be maintained.

(4) Internal stresses are set up in the metal that remain in the metal unless they are removed by proper heat treatment.

(5) Strength and hardness of the metal are increased but ductility is decreased.

(6) Recrystalline temperature for steel is increased.

(7) Resistance to corrosion is decreased while ultimate tensile strength, yield point and hardness are increased.

(8) Loss of metal due to oxidation is prevented.

Limitations

(1) The grain structure is not refined and residual stress have harmful effects on certain properties of metals.

(2) Tooling costs are high and as such it is used when large quantities of similar components are required.

(3) Only small sized components can be easily cold worked as greater forces are required for large sections. Due to large deforming forces, heavy and expensive capital equipment is required.

(4) Many of the metals having less ductility, e.g. carbon steel and certain alloy steels can not be cold worked at room temperature. It is, therefore, limited to ductile metals and range of shapes produced is not as wide as can be obtained by machining.

4.9 COLD WORKING METHODS

The principal methods of cold working are as follows

(1) Rolling

(2) Forging

(3) Drawing

(4) Bending

(5) Spinning

(6) Extruding

(7) Squeezing

(8) Peening

(9) Sizing, coining and hobbing

4.9.1 Rolling

Cold rolling is applied during the finishing stages of production of both strip and section and also in the production of very thin materials such as foils. It is also used to finish the hot rolled components to close tolerances and improve their toughness and hardness. In most cases cold rolling mills are similar in design to those used for hot rolling. The coefficient of friction for cold rolling usually ranges between 0.02 and 0.3, depending on the materials and lubricants are used. Cold rolling improves machinability by conferring the property of brittleness, a condition which is conducive to smooth tool finishes and well broken chips.

Bars of all shapes, rods, sheets, and strips are commonly finished by cold rolling.

Plates, which are generally regarded as having a thickness greater than 6 mm, are used for structural applications such as ship hulls, boilers, bridges, girders, machine structures, and nuclear vessels. Plates can be as much as 0.3 m thick for the supports for large boilers, 150 mm for reactor vessels and 110 to 125 mm for battleships and tanks.

Sheets are generally less than 6 mm thick are used for automobile bodies appliances, containers for food and beverages, and kitchen and office equipment. Commercial aircraft fuselages are usually made of about 1 mm thick aluminium alloy sheets while beverages cans are made of 0.15 mm aluminium sheet. Aluminium foil used to wrap candy and cigarette has a thickness of 0.008 mm sheets are provided as flat pieces or as strip in coils to manufacturing facilities for further processing into products.

4.9.2 Cold forging

Forging denotes a family of processes by which plastic deformation of the workpiece is carried out by compressive forces. Cold forging can be carried out at room temperature.

Proper design of forging dies and selection of die materials require considerable experience and knowledge of the strength and ductility of the workpiece material, its sensitivity to strain rate and temperature, and its frictional characteristics.

Swaging or cold forging operation consists of applying compressive or impact forces on the metal below the recrysta- llization temperature. It causes the metal to flow in the prede- termined shape according to the design of the dies. Rotary swaging and cold heading are the two important processes of swaging.

Rotary swaging is a process of reducing the diameters of rods, tubes and bars by rotating the dies which open and close rapidly on the work. This causes the rod or tube to be tapered at one end or reduced in size by combination of pressure and impact. Swaging operation is severe and it requires annealing if much reduction in area is to be carried out. The swaging operation is performed in a swaging machine which consists of a hollow spindle carrying die section as shown in Fig. 4.34.

Cold heading is another form of swaging operation and is used for the manufacture of bolts, rivets hexagonal socket screws, screws and similar headed items. The importance of cold heading operation lies principally in the saving of material, very little or no scrap being produced. Cold heading can be done on a wide variety of materials.

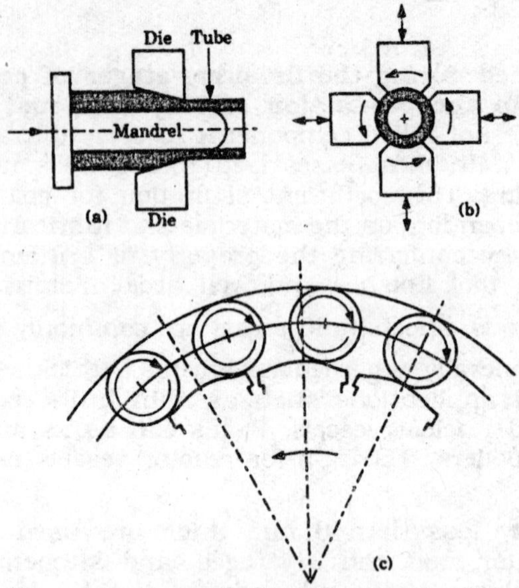

Fig. 4.34. Rotary swaging

4.9.3 Drawing (Rod, wire and tube drawing)

Drawing is an operation in which the cross sectional area of a bar or tube is reduced by pulling it through a converging die. The die opening may be any shape. Wire drawing involves smaller diameter materials than rod drawing, with sizes, as small as 0.025 mm. The drawing process, which was an established art by the 11th century, is somewhat similar to extrusion, except that in drawing the bar is under tension, whereas in extrusion it is under compression.

Rod and wire drawing are usually finishing processes. The product is either used as produced or is further processed into other shapes, usually by bending or machining. Rods are used for various application, such as small pistons, tension-carrying structural members, shafts, and spindles and as the raw material for fasteners such as bolts and screws. Wire and wire products have a wide range of applications, such as electrical and electronic equipment and wiring cables, springs, musical instruments, fencing bailing, wire baskets, and shopping carts.

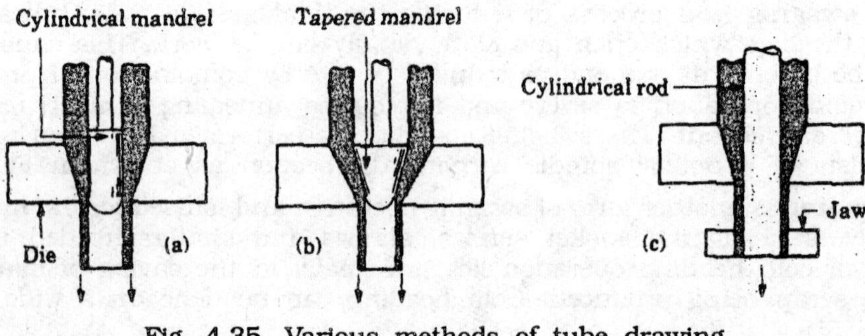

Fig. 4.35. Various methods of tube drawing.

Tubes produced by extrusion or other processes can be reduced in thickness or diameter (tube sinking) by the tube drawing processes illustrated in Fig. 4.35.

A variety of mandrels are used to produce different shapes. Shape changes can also be imparted by using dies and mandrels with various profiles.

Defects in drawing are centre cracking and the formation of seams.

Because of inhomogenous deformation, a cold drawn rod wire, or tube usually contains residual stresses. A wide range of residual stresses are present within the rod in three principal directions. However, for light reductions, the surface residual stresses are compressive light reductions are equivalent to shot peening or surface rolling thus improving fatigue life. In addition to fatigue, residual stresses can also be significant in stress-corrosion cracking over a period of time, and in warping of the component when a layer is subsequently removed, as by machining or grinding.

Die angles for successful drawing usually range from 6 degree to 15 degree. Drawing speeds depend on the material and cross sectional area. They may be as low as 0.15 m/s for heavy sections, to as high as 50 m/s for very fine wire. Temperature can rise substantially at high drawing speeds.

4.9.4 Bending

One of the most common metal working operations is **bending**. This process is used not only to form parts such as flanges, curls, seams, and corrugations but also to impart stiffness to the part by increasing its moment of inertia (note, for instance, that a long strip of metal is much less rigid when flat than when it is formed into the shape of a gutter).

Bends may be made to a sharp angle or to a radius. Sharp bends can be made in a vice, or by hammering the work over the anvil edge, but in either case the metal should be thickened by upsetting on the **outside** of the bend before the bend is started. This compensates for the distortion which takes place at the corner, the metal being compressed on the inside and stretched on the outside. The effect can be shown by bending a piece of indarubber and observing the bend produced. The metal burges wider at the inside and is stretched further at the outside, with local thinning.

Radius bends can be made over the beak of the anvil, around a bar in a vice or round a mandrel in a bending fixture. There is a again a change in cross section but the tendency is less severe, and change in width is corrected after bending.

Roll forming

For bending continuous lengths of sheet metal and for large production runs, roll forming is used. The metal strip is bent in stages by passing it through a series of rolls (Fig 4.36). Typical products are channels, gutters, siding, panels, frames, and pipes and tubing with lock seams. The length of the part is limited only by the amount of material supplied from the coiled stock. The parts are usually sheared and stocked continuously. The sheet thickness typically ranges from about 0.125 mm to 20 mm. Forming speeds are generally below 1.5 m/s, although they can be much higher for special applications.

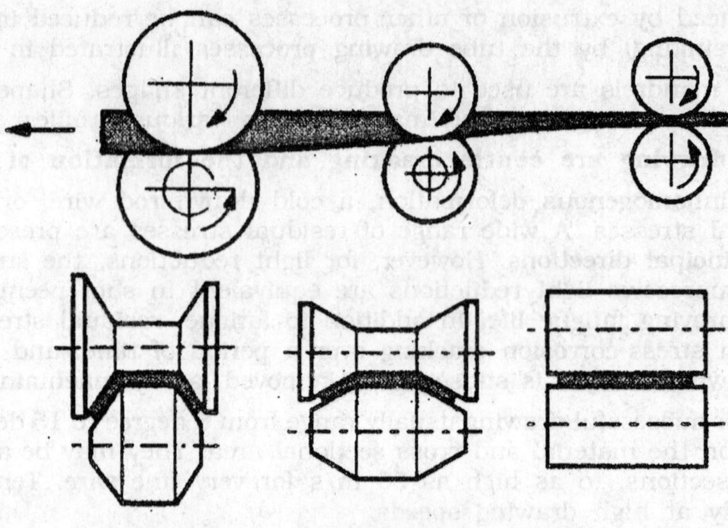

Fig. 4.36. The roll forming process.

4.9.5 Spinning

In the spinning, an object with surface of revolution is produced from a sheet metal. The blank is held against a form die which is rotated and the sheet metal blank is laid over this die using a specially-shaped tool or roller. If a simultaneous thinning of the sheet metal takes place during the operation, the process then is called shear spinning. Both these operations are shown in Fig. 4.37.

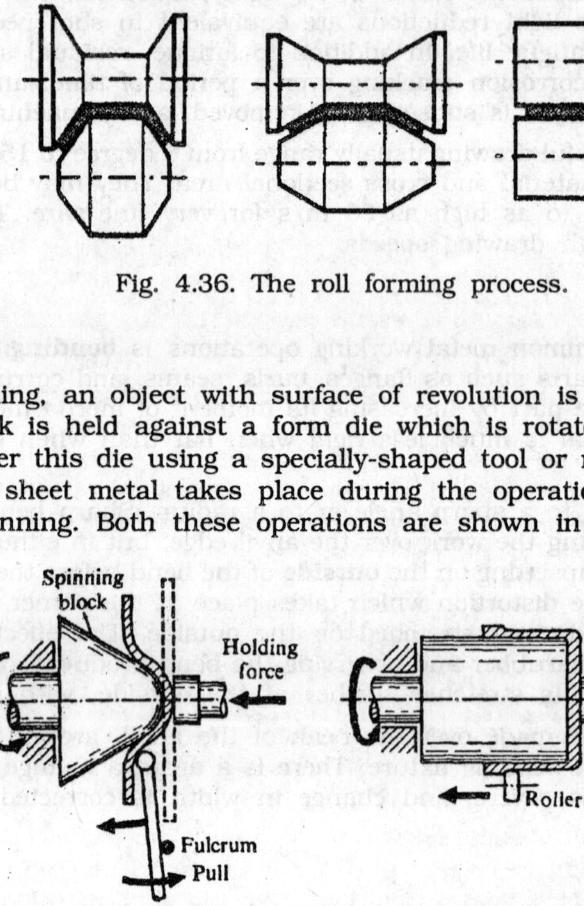

Fig. 4.37. Spinning and shear spinning operatins.

4.9.6 Extruding

Cold extrusion is general term often used to denote a combination of processes, such as direct and indirect extrusion and forging (Fig 4.38).

Many materials can be extruded into various configurations, with the billet either at room temperature or at a temperature of a few hundred degrees.

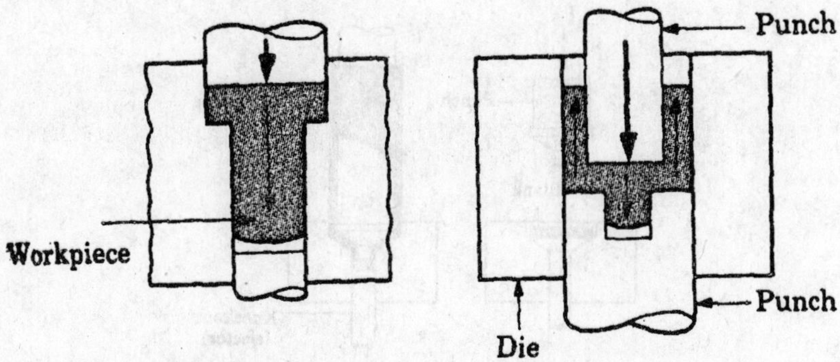

Fig. 4.38. Cold extrusion- an example.

Cold extrusion has gained wide acceptance in industry because of the following advantages over hot extrusion.

(1) Improved mechanical properties resulting from strain hardening, provided that the heat generated by plastic deformation and friction does not recrystallize the extruded metal.

(2) Good control of tolerances, thus requiring a minimum of machining operations.

(3) Improved surface finish, provided that lubrication is effective.

(4) Lack of oxide layers.

(5) High production rates and relatively low cost.

However, the stresses on tooling in cold extrusion are very high, especially with steel workpieces. The design of tooling and selection of appropriate tool materials are therefore crucial to success in cold extrusion.

Impact extrusion

A process often included in the category of cold extrusion, impact extrusion, is similar to indirect extrusion. The punch descends at high speed and strikes the blank (slug), extruding it upward. The thickness of the extruded tubular section is a function of the clearance between the punch and the die cavity. The process already discussed earlier.

The impact extrusion process usually produces tubular sections having wall thicknesses that are small in relation to their diameters. This ratio can be as small as 0.005. The concentricity between the punch and the blank is important for uniform wall thickness. A typical example of impact extrusion is the production of collapsible tubes, such as for toothpaste (Fig 4.39). The punch travel is determined by the setting of the press.

4.9.7 Squeezing

Squeezing is a quick and widely used way of forming ductile metals. Of the different processes of squeezing, rotary swaging or cold forging and cold heading are very common. Both the processes are explained in the heading of cold forging.

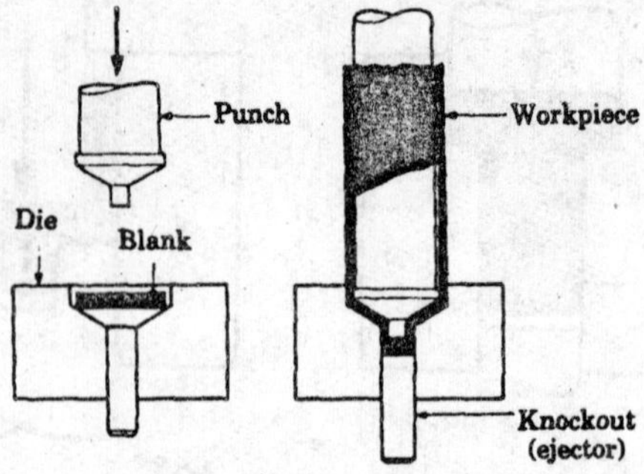

Fig. 4.39. Impact extrusion of a collapsible tube (Hooker process)

4.9.8 Peening

This method is employed to set up a superficial state of surface compressive stress, causing the interior of the member to assume an opposite tensile stress. Because fatigue generally occurs from surface cyclically loaded in tension, the useful lives of such members are frequently extended by shot peening. This method is sometimes employed to achieve an ornamental effect.

Shot peening improve the fatigue resistance of metal by setting up compressive stresses in the surface. The process consists of throwing a blast of metal shots on to the surface of a component. Shot peening is done by air blast or by some mechanical means. Peen forming is used by the air craft industry to generate smooth and couple curvatures on aircraft wing skins.

4.9.9 Sizing, coining and hobbing

Malleable iron, forged steel, powdered metals, aluminium and other ductile nonferrous metals parts are commonly finished to thickness by squeezing an operation call **sizing**. A special die is needed for almost every job, but each piece can be sized in a fraction of the time of machine. Thus, sizing is economical wherever applicable in high production industries.

Coins, metals and other similar parts are produce by **coining**. It requires the use of die and punch. which are of robust structure as high pressure is required during the operation. The dies used are such that they confine the metal and restrict its flow in the lateral direction. Since high pressure is required in the operation, its use is limited to fairly soft alloys.

Hobbing is a process of producing cavities into softer metals by forcing a hardened steel from a hob. This operation is performed on heavy presses of capacities ranging from 250 to 8000 tonnes. During the operation the flow of metal in the blank is restrained from lateral movement by the heavy retainer ring placed around it.

4.10 FORMABILITY OF SHEET METALS

The formability of sheet metals has been of great interest because of its technological and economic significance. Sheet metal formability is generally defined as the ability of the metal to undergo the desired shape changes without failures such as necking and tearing. Three factors have a major influence on formability.

(1) Properties of sheet metal,

(2) lubrication at various interfaces between the sheet and dies and tooling and

(3) the characteristics of the equipment and tools and dies used in forming the sheet.

4.11 FORGEABILITY

The forgeability of a metal can be defined as its capability to undergo deformation by forging without cracking. This definition can be expanded to include the flow strength of the metal. Thus a material with good forgeability is one that can be shaped with low forces without cracking.

4.12 METALLURGICAL ADVANTAGES OF HOT WORKING OVER COLD WORKING PROCESSES

From the metallurgical point of view the hot working have greater advantage over cold working processes. The metallurgical advantages of hot working over cold working processes are given in table 4.1.

Table 4.1 Metallurgical advantages of hot working over cold working processes.

SN	Hot Working	Cold working
1.	Internal or residual stresses are not developed in the metal.	Internal and residual stresses are developed in the metal.
2.	Cracks and blow holes are welded up.	Possibility of crack formation and it propogation is great.
3.	Refinement of crystals occurs.	Crystallisation does not occurs. Grains are only elongated and/or distorted. The extend of distortion depend on the degree of cold working.
4	Break up brittle film of hard constituents, and promotes uniformly of materials by facilitating diffusion of constituent alloys.	Uniformly of material is lost and the properties are affected.
5	Hot working is done at a temperature above recrystalli zation temperature. It can therefore be regarded as a simultaneous process of deformation & recovery.	Cold-working is done at a temperature below recrystallization temperature. So no appreciable recovery can take place during deformation.
6	Mechanical properties such as elongation, reduction of area and impact values are improved. Ultimate tensile strength, yield point, fatigue strength, hardness, and resistance to corrosion, are not affected by hot working if this is properly done.	Cold working decreases elongation, reduction of area, and impact values, while it increases ultimate tensile strength, yield point, fatigue strength and hardness. But resistance to corrosion is decreased by cold working. If severely worked, yield point may coincide with ultimate tensile strength
7	Hardening due to plastic deformation is completely eliminated by recovery and recrystallization. This is happen, however, only if the rate of crystalization is higher than that of deformation.	Hardening is not eliminated since working is done at a temperature below recrystallization, and this is always accompanied by work hardening.

EXERCISES

A. Theory

4.1 What do you understand by mechanical working of metals?

4.2 Define hot working of metals. What are its advantages and disadvantages?

4.3 Name the various methods of hot working and explain any two.

4.4. Describe the different types of rolling mills used for rolling.

4.5. Explain how a cast structure is converted into a wrought structure by hot rolling.

4.6 Explain how you would go about applying front and back tensions to sheet metals during rolling.

4.7 Describe the methods by which roll flattening can be reduced.

4.8. Surface roughness in hot rolled products is higher than in cold rolled products. Explain why?

4.9 Explain the advantages and limitations of using small diameter rolls on flat rolling.

4.10 Explain the working and application of hot and cold rolling with the help of neat sketches.

4.11 Describe the process of hot rolling of steel bars.

4.12 Explain the principle of rolling. How are rolling mills classified?

4.13 How can you tell whether a certain point is forged or cast? Describe the features that you would investigate to arrive at a conclusion.

4.14 Define forging. Explain open and closed die forging with a neat sketch.

4.15 Give the temperature range for forging in terms of the homologous temperature.

4.16 Draw a neat sketch for the effect of recovery, recrystallization and grain growth on mechanical properties, shape and size of grains.

4.17 Define recrystallization temperature. Explain how the final hot working temperature influence the properties of the metal.

4.18 Define forging what are the advantages and disadvantage of forging.

4.19 Name the various methods of forging and explain any two.

4.20 Draw a neat sketch showing the principle of rotary forging.

4.21 Draw a neat sketch showing the principle of roll forging.

4.22 Draw a neat sketches of closed upsetting and open upsetting and explain their operation.

4.23 Explain the not working process of piercing with a neat sketch.

4.24 Explain the principle of drawing with a neat sketch.

4.25 Explain the principle of hot spinning with a neat sketch.

4.26 Draw neat sketches of the four basic types of extrusion.

4.27 Explain the principle of backward extrusion with a neat sketch.

4.28 Explain the method of extruding tubular section.

4.29 Explain the cold working impact extrusion process.

4.30 Draw neat sketches of various extrusion processes.

4.31 Define cold working of metals.

4.32 Explain the principle of cold working.

4.33 What are the advantages and limitations of cold working.

4.34 Name the various methods of cold working of metals explain any two.

4.35 Explain the cold rolling process.

4.36 Explain the cold forging process.

4.37 Draw a neat sketch of rotary swaging and explain.

4.38 Explain cold drawing process.

4.39 Draw various methods of tube drawing.

4.40 Explain cold bending process

4.41 Explain roll forming process with a neat sketch.

4.42 Draw neat sketch of spinning and shear spinning operations.

4.43 Explain cold extruding.

4.44 Differentiate between hot extrusion and cold extrusion.

4.45 Explain cold squeezing process

4.46 Explain cold peening process.

4.47 Explain sizing, coining and hobbing.

4.48 Define formability of sheet metals. Explain the three major factors influences on formability.

4.49 Define forgeability. On what factors does it depend?

4.50 Explain forgeability of various metals.

4.51 Discuss the metallurgical advantages of hot working over cold working processes.

4.52 The extrusion ratio, die geometry, extrusion speed, and billet temperature all affect the extrusion pressure. Explain why?

4.53 Explain why the die pressure in drawing decreases toward the exit.

4.54 We have seen that in rod and wire drawing the maximum die pressure is at the die entry. why?

4.55 Many missile components are made by spinning what other methods would you use if spinning processes were not available?

4.56. Explain (i) Bending (ii) Drawing. **[RU 2002]**

4.57. Differentiate between hot working and cold working process. **[RU 2001]**

4.58. (a) Give a brief comparison of hot working and cold working.
(b) Explain the following forging operations with the help of neat sketches (i) upsetting (ii) drawing down (iii) setting down (iv) fullering and (v) cutting. **[MREC, 2000]**

4.59. (a) What do you understand by the recrystallisation and recrystallisation temperature?
(b) How is the grain structure of the metal affected during rolling?
(c) What is hot extrusion? Explain extrusion processes with the help of neat sketches. **[RU 1999]**

4.60. (a) Explain with the help of neat sketches:
(i) Extrusion
(ii) Spinning
(b) What is drop forging? How is it done?
(c) Describe how an ingot is converted into a bar stock by the process of rolling. **[RU 1998]**

4.61. Write short nots on "Bending and forming Operation" **[RU 1998]**

4.62. Describe the tube drawing operation with the help of a suitable sketch. **[RU 1997]**

4.63. Give a brief description of the following manufacturing processes with sketches: (i) Hot and cold rolling, (ii) Forging, (iii) Shearing, (iv) Drawing, (v) Rivetting, (vi) Cold heading, (vii) Bending & (viii) Forming **[RU 1996]**

4.64. (a) Classify the hot working processes. Explain the working and application of hot rolling and forging with the help of neat sketches.
(b) What are the advantages and limitations of cold working process? **[RU 1995]**

5

FOUNDRY

5.1. INTRODUCTION

Foundry industry bases itself on the founding or casting principles. **Foundry is defined as a collection of materials, fuel, fluxes and tool to produce castings.**

Casting is one of the oldest manufacturing process, and even today is the first step in manufacturing most products. In this process, the material is first liquefied by properly heating it in a suitable furnace. Then, the liquid is poured into a previously prepared mould cavity where it is allowed to solidify subsequently the product is taken out of the mould cavity, trimmed, and cleaned to shape.

It is clear from the definition of the process that a successful casting operating needs a knowledge in the following areas :

(i) Preparation of moulds and patterns (used to make the mould).

(ii) Melting and pouring of the liquefied metal.

(iii) Solidification and further cooling to room temperature.

(iv) Defects and inspection.

There are various types of casting processes depending, among others, on the material, the type of patterns and moulds, and the pouring technique.

The suitability of the casting operation for a given material depends on

(i) the melting temperature of the job and the mould materials.

(ii) the solubility of and the chemical reaction between the job and the mould materials,

(iii) the solubility of the atmosphere in the material at different temperatures to be encountered in the casting operation,

(iv) the thermal properties such as conductivity and coefficient of linear expansion of both the mould and job materials.

Fig. 5.1 shows the flow chart of a typical foundry operation.

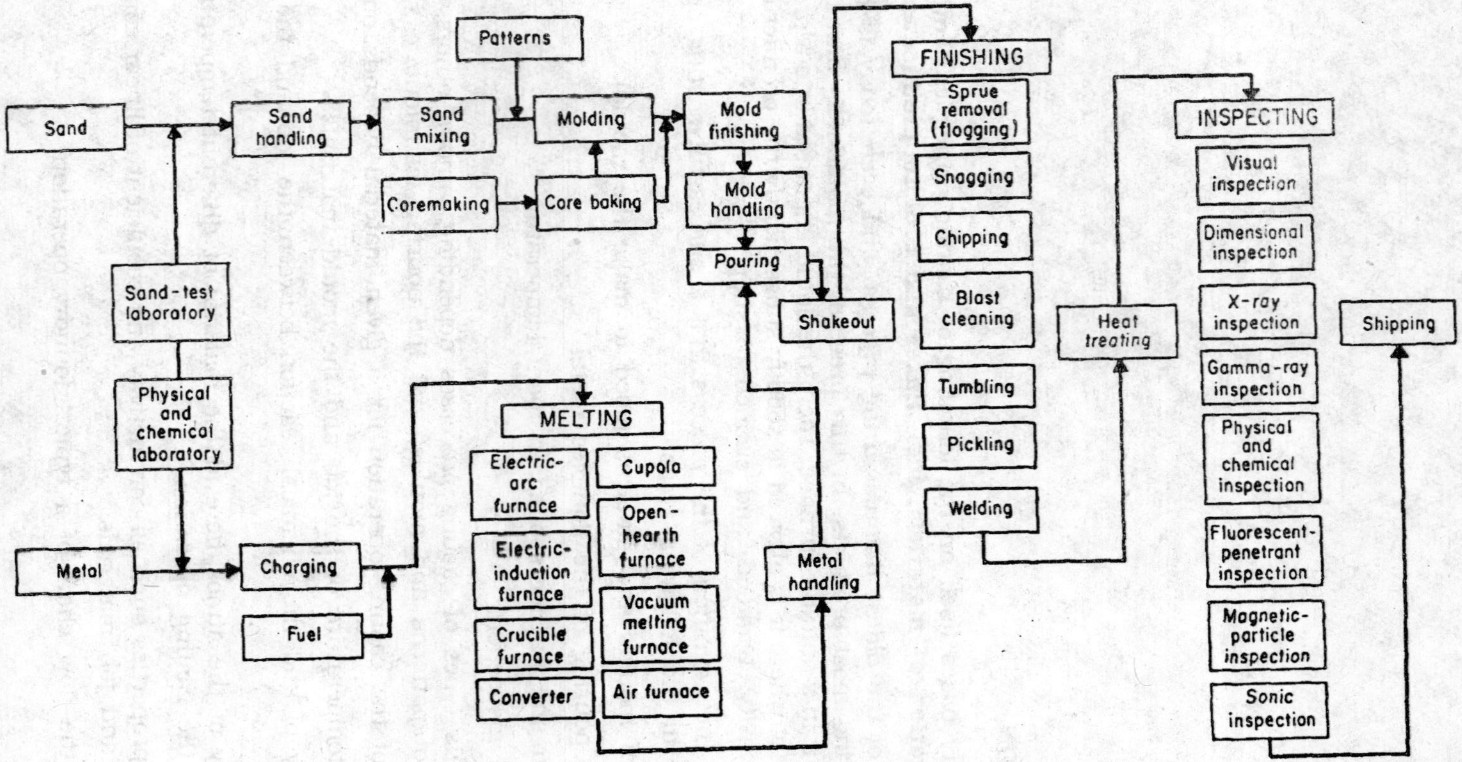

Fig. 5.1. Flow chart of a typical foundry operation.

5.2. MOULD MAKING

Moulds are made by hand or machine. Hand moulding is slow and is used for job works, where one or few castings are made. Machine moulding is used for large scale or mass production.

5.2.1. Foundry tools and equipment

A variety of hand tools and equipment are needed for performing different operations like lifting, shifting, casting, machining, and inspecting. Depending upon the size, shape and number of castings to be produced and nature of work to be done, the foundry tools and equipment can be classified into the following categories.

1. Pattern making tools.
2. Core making tools.
3. Sand conditioning and testing equipments.
4. Mould making tools.
5. Melting and pouring equipment.
6. Cleaning and finishing equipment.
7. Testing equipment.

1. Hand tools : Various hand tools are used in a foundry for making moulds. A brief description of various hand tools are gives below.

a. Shovel : It consists of a square iron blade fitted with a D- shaped wooden handle. It is used for transferring sand to the sand mixer, and pouring sand in the moulding flask.

b. Strike off bar : It is a wooden or metallic bar having a true edge. It is used for removing the surplus sand after ramming has been completed.

c. Hand riddle : It is also known as a sieve. It consists of a wooden frame fitted with a screen of standard wire mesh at its bottom. It is used for removing foreign materials such as nails, shot metals, splinters of wood, etc. from the moulding sand. Different types of riddles are denoted by different numbers like 8, 10, 12, etc. Riddle no. 8 has an opening of 1/8 inch approximately.

d. Vent wire : It is a thin steel rod or wire having a pointed edge at one end and a wooden handle or a bent loop at the other end. It is used for making small holes, called vents in the rammed sand moulds to permit easy escape of gases and steam generated during cooling of the heated metal.

e. Trowels : Trowels are made of iron and are provided with a wooden handle. The three types of commonly used trowels are

(i) Finishing trowel. (ii) Square end trowel. (iii) Heart shaped trowel.

Trowels are used for

(i) Finishing flat surfaces for moulds.

(ii) Cutting ingates.

(iii) Making joints.

(iv) Repairing moulds.

(v) Giving shape.

f. Slicks : Slicks are used for repairing and providing finishing to mould surface and edges after the pattern has been withdrawn. The commonly used slides in a foundry are shown.

g. Lifter (cleaner) : The lifters as shown are made of thin sections of steel of various width and lengths with one end bent at right angles. The lifter is a finishing tool used for patching deep sections of a mould and removing loose sand from pockets of the moulds. The Yankee lifter is very commonly used in an average sized foundry.

h. Rammers : Rammers are used for packing and ramming the sand in the moulding box. Depending upon the type of casting and nature of working there are many types of rammers. Rammers are generally made of wood or steel. The commonly used rammers.

An automatic pneumatic rammer operated by compressed air. It is fitted with long or short shafts according to the requirements with open or flat ends. It is used for ramming sand in large moulds. It saves considerable labour and time in moulding.

i. Swab : It is a small brush having long hemp fibres. A swab is used for applying water around the edges of a pattern. This prevents the sand edges from crumbling when the pattern is removed from the mould. A bulb swab has a rubber bulb to hold the water and a soft hair brush at the open end.

j. Sprue pin : A sprue pin is also known as a rubber peg. It is a tappered wooden or iron rod, which when embedded into top part of the mould known as cope and later withdrawn produces a cavity known as down gate or runner. Through the runner the molten metal is powered into the mould.

k. Draw spike : It consists of a tappered steel rod having a loop or ring at one end and a sharp point at the other. It is used to draw a pattern from the mould.

l. Draw screw and rapping plate : The draw screw is a straight mild steel rod carrying a loop or ring at one end and a wood or machine screw at the other. It's always used in conjunction with a rapping plate for rapping and withdrawing the pattern from the sand. The rapping plate or lifting plate is provided with several threaded and rapping holes. Threaded holes accommo- date either a wood screw type or machine screw type draw screws. Rapping holes are provided to accommodate separate rapping rods so that the threaded holes are not damaged.

m. Smoothers and corner slicks : These are tools used for giving a finish to and repairing of

(i) round and square corners and edges

(ii) round and flat surfaces.

According to their use and shape, they are given names. Most commonly used tools for giving finishing and repairing are inside square, flat, pipe, button, half round corner and egg shaped.

n. Clamps : Clamps are made of steel and are used for holding and clamping the moulding boxes in proper position. Commonly used clamps.

2. Moulding boxes or flasks : Various types of moulding boxes are used in the foundry

(i) Box type flash

(ii) Tapered slip flask

(iii) Snap flask

(iv) Wooden moulding box.

A moulding box is a container that holds the rammed sand rigidly and allows the molten metal to solidity after casting in a mould cavity.

Moulding boxes are generally made in two parts. The top part is called the cope and bottom is called the drag. Moulding boxes are made of wood, cast iron or steel. Wooden boxes are used for limited production. Boxes made of fabricated steel are light, robust and can withstand impact. In general a moulding box must be capable of withstanding rough handling.

Selection of moulding box regarding its shape and size mainly depends upon the type of product to be cast. Since it is not possible to make as many moulding boxes as the number of components, it is always better to standardise the size of moulding boxes.

3. Ladles : Various type of ladles used in foundries are

(i) Holding ladle.

(ii) Hand or shake ladle

(iii) Bull ladle

(iv) Crane or monorail ladle

(v) Teapot ladle

(vi) Bottom stoppered ladle.

Ladles are used to transport molten metal from the melting furnace to the mould and vice versa. The molten metal is often held temporarily in large holding ladles, from which it is tapped when needed. These holding ladles consist of a steel shell lined with a suitable refractory material like fireclay.

4. Crucibles : The shape of crucibles is similar to ladles. They are made of refractory material and are used as metal melting pots. The charge in the form of small pieces is kept in crucibles. The main disadvantages of crucibles is that they cannot hold a high amount of charge and control of chemical composition is also difficult.

5.2.2 The process of mould making

(1) Preparation of drag : Place the patten on the moulding board. Sprinkle some parting sand (clean, dry river sand) to prevent sticking of foundary sand to the moulding board.

Place the drag with its dowel pins facing downwards and fill the box with foundry sand.

Ram the sand initially with the peen and ensure that the sand covers the pattern thoroughly. Then fill the box completely with sand, ram and level up the top with a strike off bar. Vent the sand packing using vent wire.

(2) Preparing cope :- Now roll over the drag on to a bottom board (similar to moulding board) and assemble the cope with the lugs engaging the dowels on the drag. Sprinkle parting sand inside to prevent cope and drag clinging to each other. With sprue pin and riser pink placed in position, fill the cope with foundary sand, ram and vent.

Slowly the cope from the dag, remove the pattern carefully with the help of draw spike, cut out a gate connecting the bottom sprue passage with the mould cavity.

(3) Reassemble the cope and drag. The mould is now ready for casting.

Usually the mould surfaces is given a thin coating of graphite mixture using a brush. This improves the surface finish of the casting.

5.2.3. Moulding materials

A large variety of moulding materials is used in foundries for manufacturing moulds and cores. These are as follows :

(a) Moulding sand.

(b) Backing sand.

(c) Facing sand.

(d) Parting sand

(e) Core sand.

The choice of moulding materials is based on their processing properties.

Refractoriness : It is the ability of the moulding material to withstand the high temperatures of the molten metal so that it does not cause fusion.

Green strength : The moulding sand that contains moisture is termed as green sand. The green sand should have enough strength so that the constructed mould retains its shape.

Dry strength : When the moisture in the moulding sand is completely expelled, it is called dry sand.

Hot strength : After all the moisture is eliminated, the sand would reach a high temperature when the metal in the mould is still in the liquid state. The strength of the sand that is required to hold the shape of the mould cavity then, called hot strength.

Permeability : During the solidification of a casting, large amounts of gases are to be expelled from the mould. The gases are those which have been absorbed by the metal in the furnace, air absorbed from the atmosphere and steam and other gases that are generated by the moulding and core sands. If these gases are not allowed to escape from the mould, they would be trapped inside the casting and cause defects. The moulding sand should be sufficiently porous so that the gases are allowed to escape from the mould. This gas evolution capability of the moulding sand is termed as permeability.

Besides these specific properties, the moulding sand should also have collapsibility so that during the contraction of the solidified casting, it does not provide any resistance which may result in cracks in the casting they should be re-usable and should have good thermal conductivity so that heat from the casting is quickly transferred.

A mould material is the raw material for making moulds. It should be such that a casting can be made of it and its cavity is capable of retaining its shape till the molten metal has solidified. Moulds can be classified into two categories.

(a) Temporary moulds.

(b) Permanent moulds.

a. Temporary moulds : Temporary moulds are made of refractory sand and other binding materials. These are suitable for all types of ferrous castings, i.e. cast iron, wrought iron, steel etc. Its main drawback is that it is suitable for one casting only and it has to be made afresh for each casting with the help of patterns.

b. Permanent moulds : Permanent moulds are generally made up of ferrous metals and alloys, i.e. cast iron, steel, etc. Permanent moulds are suitable for casting low melting point alloys. Since permanent moulds are expensive they are suited for large scale production of similar components of small size. Due to high cost of permanent moulds, most of the castings in a foundry are done in temporary moulds of refractory sand.

In comparison to permanent sand moulds, temporary sand moulds are suitable for high melting point materials and larger objects permanent moulds are suitable for low melting alloys and small sized components with greater dimensional accuracy.

Fig. 5.2 shows a typical mould ready for pouring. The upper portion is called the "cope", and the lower portion is termed the "drag". A sand mould is ordinarily made up of two or more bodies of sand which separate in order to withdraw the pattern. They are then placed together again to produce a mould cavity of the desired

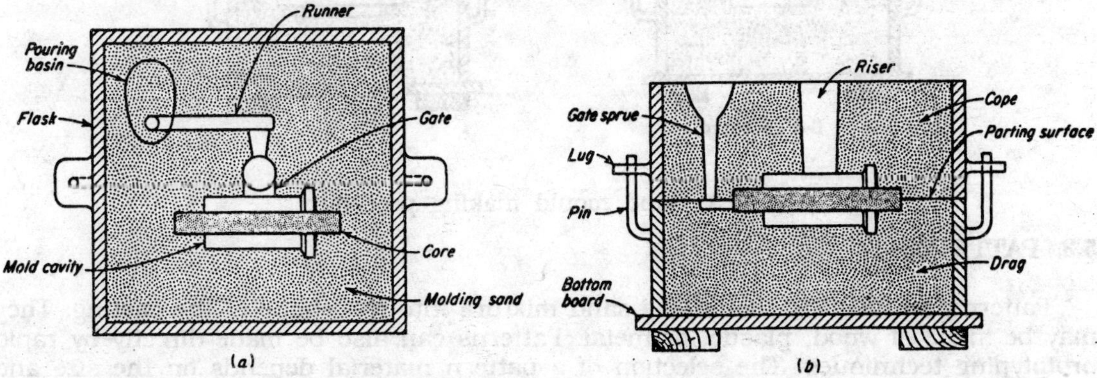

Fig. 5.2. Section views through a completed sand mould ready for pouring (a) Top view (b) Front view.

shape. The boxlike frame, with no top or bottom, into which the sand is rammed is known as a "flask". In this case it consists of two parts, one for the cope and the other to accommodate the drag. Pins and lugs with holes which fit together ensure proper alingment during the replacement of cope upon drag. A parting surface, or joint, is that surface where two portions of a mould, such as the cope and drag, meet. Sometimes vent holes may be formed in a mould to help the escape of steam and other gases produced during pouring. However, this is usually not required because a sand mould properly rammed should ordinarily have sufficient spaces between the sand grains to allow these grains to escape. A core which has sufficient strength for handling is placed in the mould cavity to form the hole through this casting. Molten metal will be pointed into the pouring basin. From there it will flow downward through the gate sprue to a runner and enter the mould cavity through the gate or gates. A riser is provided for this casting to supply molten metal to compensate for the liquid metal shrinkage during the solidification of the casting. Gates and risers are removed from the castings in the cleaning department and later remelted.

Fig. 5.3 shows the sand mould making procedure.

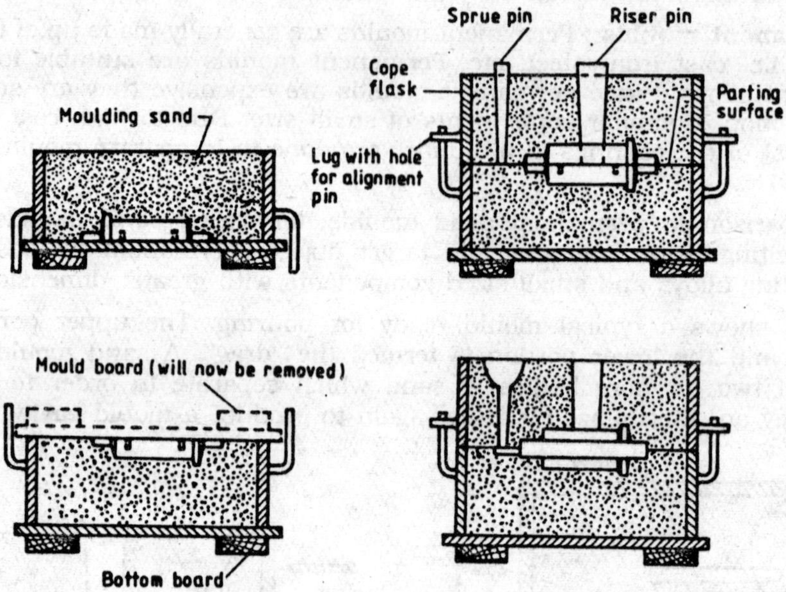

Fig. 5.3. Sand mould making procedure.

5.3. PATTERNS

Patterns are used to mould the sand mixture into the shape of the casting. They may be made of wood, plastic, or metal. Patterns can also be made directly by rapid prototyping techniques. The selection of a pattern material depends on the size and shape of the casting, the dimensional accuracy and quantity of castings required, and the moulding process to be used. Because patterns are used repeatedly to make moulds, the strength and durability of the material selected for patterns must reflect the number of castings the mould will produce. Patterns may be made of a combination of materials to reduce wear in critical regions. Patterns are usually coated with a parting agent to facilitate their removal from the moulds.

5.3.1. Difference between pattern and casting

A pattern differs from actual component in the following ways.

(1) It carries various shrinkage allowance.

(2) It carries draft allowances.

(3) It carries core prints.

(4) It gives a machine allowance to clean and finish the required surfaces.

(5) Patterns may be in two or three pieces whereas castings are in single piece.

(6) Patterns are made of different materials like wood, cast iron, aluminium, brass etc.

5.3.2. Functions of a pattern

A pattern serves the following functions

(1) A properly manufactured pattern having a good surface finish reduce casting defects.

(2) A pattern prepares a mould cavity to be used as a casting.

(3) Risers, runners and gates may form a part of the pattern.

(4) Properly constructed patterns reduce the overall cost of a casting.

(5) A pattern may contain core prints for placing the cores in proper positions.

(6) A pattern helps in establishing a parting line between the cope and drag part of the mould.

(7) A pattern helps in positioning the core at its proper position.

5.3.3. Pattern materials

Patterns may be constructed from the different materials with respect to their own advantages, limitations and the field of applications. The selection of pattern materials depends upon the following factors :

(1) Method of moulding i.e. hand or machine moulding.

(2) Shape, complexity and size of the casting.

(3) Types of moulding materials i.e. sand etc.

(4) The desired dimensional accuracy and surface finish required for the casting.

(5) Casting design parameters (i.e. minimum section thickness) and the complexity of the cast part.

(6) Nature of the moulding process.

(7) The number of castings to be produced. Metal pattern are preferred when the production quantity is large.

(8) The chances of repeat orders.

(9) The high probability of changing the casting and hence the pattern in near future.

Different materials used for making patterns are listed below

(1) Wood

(2) Metal

(3) Plastic

(4) Plaster

(5) Wax

(6) Polystryrene.

(1) Wood

The most common materials for making patterns for sand casting is the wood.

Advantages

(1) Inexpensive.

(2) Easily available in large quantities.

(3) Light in weight.

(4) Easy to join to acquire complex and large pattern shapes.

(5) Easy to machine and shape to different configurations and forms.

(6) Easy to obtain good surface finish.

(7) Repaired easily.

Limitations

(1) Weak as compared to metal patterns.

(2) Cannot withstand rough handling.

(3) Abraded easily by sand action.

(4) Posses poor wear resistance.

(5) Susceptible to shrinkage and swelling.

(6) Absorb moisture, consequently get warped and change shape and size.

Wooden patterns are used where the number of casting to be produced is small and the pattern size is large.

Wood to be used for making patterns should be free from knots and other common defects. It should be straight grained and properly seasoned.

(2) Metal

Metal patterns are cast from wooden patterns. The different metals and alloys used for making patterns are

(1) Steel

(2) Cast iron

(3) Brass

(4) White metal

(5) Aluminium and aluminium alloys.

Advantages

(1) Withstand rough handling.

(2) Greater resistance to abrasion.

(3) Do not wrap.

(4) More stronger and accurate as compared to wooden patterns.

(5) Life much longer than wooden patterns.

(6) Do not absorb moisture retain their shape.

(7) Accurate dimensional tolerances.

(8) Easy to obtain surface finish.

(9) Excellent wear resistance and strength to weight ratio.

(10) Good machinability characteristic.

(11) Stable under different environments.

(12) Afford excellent to fair resistance to corrosion and excellent resistance to swelling.

Limitations

(1) Cannot machined so easily as wooden ones.

(2) Ferrous patterns are rusted.

(3) Easily not repaired. (i.e. Aluminium patterns)

(4) Expensive as compared to wood patterns.

(5) Ferrous patterns are heavier than wooden patterns.

Metal patterns are employed where large number of castings have to be produced from the same pattern.

(3) Plastic

Advantages

(1) Light weight

(2) Good strength

(3) Wear and corrosion resistant.

(4) Provides good surface finish.

(5) Durable

(6) Provides smooth surface.

(7) Moisture resistant.

(8) Does not involve any appreciable change in its size or shape.

(9) Better adhesive quantities.

(10) Good resistance to chemical attack.

(11) Better impact strength than wood.

(12) Easy to make.

(13) Good compressive and flexural strength.

(14) Low solid strinkage.

(15) Abrasion resistant.

(16) Does not stick with mould material and can be easily taken out without spoiling the mould cavity.

Limitations

(1) Plastic patterns are fragile and thus light sections may need metal reinforcements.

(2) Plastic patterns may not work well when subject to conditions of severe shock as in machine moulding (jolting).

(4) Plaster

Advantages

(1) Easily worked by using wood working tools.

(2) High compressive strength (upto 280 bar)

(3) Unlike metals, it expands while solidifying.

(4) Intricate shapes can be cast without any difficulty.

(5) In case of a plaster (pattern) having suitable coefficient of expansion, no shrinkage allowance may be given on the pattern.

Plaster is used for making small and intricate patterns and core boxes.

(5) Wax

Advantages

(1) Provide very good surface finish.

(2) Impart high accuracy to the castings.

(3) After being moulded, the wax pattern is not taken out of the mould like other patterns; rather the mould is inverted and heated; the molten wax comes out and/or

is evaporated. Thus there is no chance of the mould cavity getting damaged while removing the pattern.

Wax patterns are made in a water cooled two piece metal mould or a die. Wax patterns find applications in investment casting process.

(6) Polystyrene

Polystyrene pattern, which evaporates upon contact with molten metal to form a cavity for the casting. The process is known as expandable pattern casting. The process is also known as evaporative pattern or lost pattern casting and under the trade name full mould process. It is a important casting processes for ferrous and nonferrous metals, particularly for the automotive industry.

Advantages

(1) It is a relatively simple process because there are no parting lines, cores, or riser systems; hence it has a design flexibility.

(2) Inexpensive flasks are sufficient for the process.

(3) Polystyrene is inexpensive and can be easily processed into patterns having very complex shapes, various sizes, and fine surface detail.

(4) The casting requires minimum finishing and cleaning operations.

(5) The process is economical for long production runs. A major fact is the cost to produce the die for expanding the polystyrene beads to make the pattern.

(6) The process can be automated.

Typical applications for this process are cylindrical heads, crankshafts, brake components and manifolds to automobiles, and machine bases.

Recent developments include the use of polymethylmethacrylate (PMMA) and polyalkylene carbonate as pattern materials for ferrous castings.

5.3.4. Pattern allowances

The following pattern allowances are provided to patterns due to which a pattern is larger in size compared to the final casting.

(1) Shrinkage allowance.

(2) Machining or finishing allowance.

(3) Draft or taper allowance.

(4) Camber or distortion allowance.

(5) Shake or rapping allowance.

(6) Mould wall movement allowance.

5.3.5. Pattern types

There are various types of patterns depending upon the complexity of the job, the number of castings required and the moulding procedure adopted.

(1) Single piece pattern

(2) Two piece or split pattern.

(3) Cope and drag pattern.

(4) Match plate pattern.

(5) Loose piece pattern.

(6) Follow board pattern.

(7) Sweep pattern.

(8) Skeleton pattern.

(9) Gated pattern.

(10) Segmental pattern.

1. Single piece pattern : These are inexpensive and the simplest type of patterns. As the name suggests the pattern is made from one piece and does not contain loose pieces or joints. Stuffing box of steam engine may be cast with the help of a one piece pattern. Fig. 5.4 shows a single piece pattern.

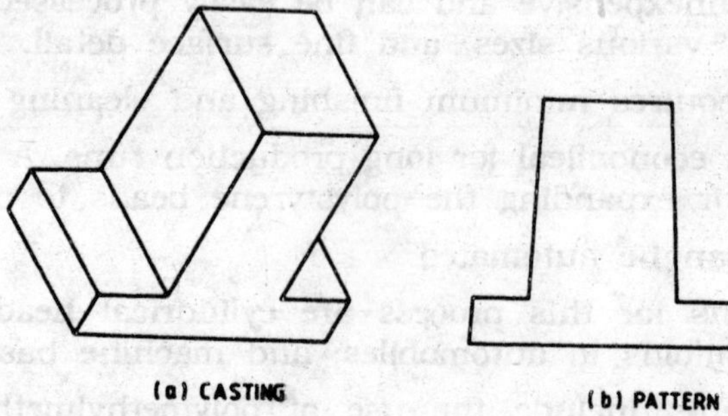

Fig. 5.4. Single piece pattern.

2. Split pattern : This is the most widely used type of pattern for intricate castings. When the contour of the casting makes its withdrawal from the mould difficult, or when the depth of the casting is too high, then the pattern is split into two parts so that one part is in the drag and the other in the cope. The split surface of the pattern is same as the parting plane of the mould. The two halves of the pattern should be aligned properly by making the use of the dowel pins which are fitted to the cope half. These dowel pins match with the precisely made holes in the drag half of the pattern and thus align the two halves properly as seen in Fig. 5.5.

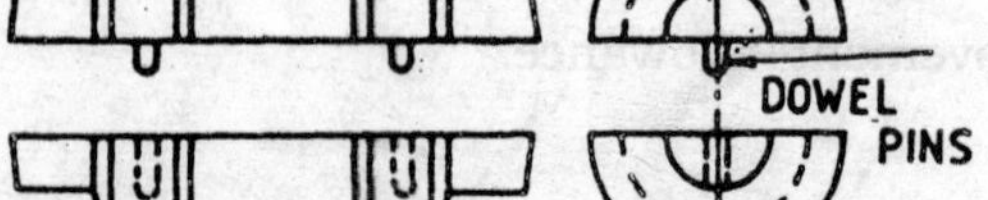

Fig. 5.5. Split pattern.

3. Cope and drag pattern : These are similar to split patterns. In addition to splitting the pattern, the cope and drag halves of the pattern along with the gatting and risering systems are attached separately to the metal or wooden plates along with the alignment pins (Fig. 5.6). They are called cope and drag patterns. The cope and drag moulds may be produced using these patterns separately by two moulders but they can be assembled to form a complete mould. These types of patterns are used for castings which are heavy and inconvenient for handling and also for continuous production.

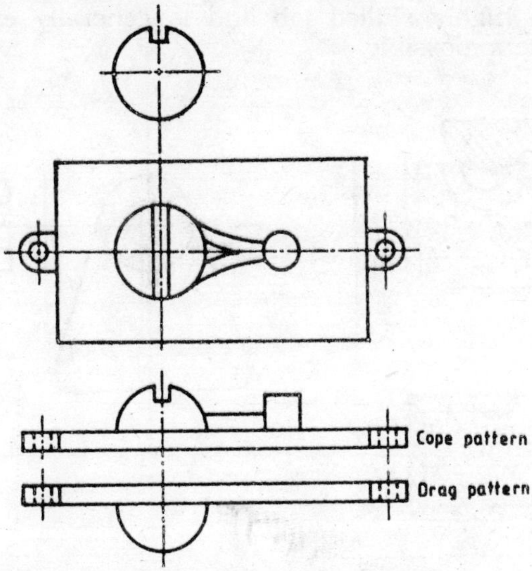

Fig. 5.6. Cope and drag pattern.

4. Match plate pattern : These are extension of the previous type. Here the cope and drag patterns along with the gating and risering are mounted on a single matching metal or wooden plate on either side as shown in Fig. 5.7. On one side of the match plate the cope flask is prepared and on other the drag flask. After moulding when the match plate is removed, a complete mould with gating is obtained by joining the cope and the drag together.

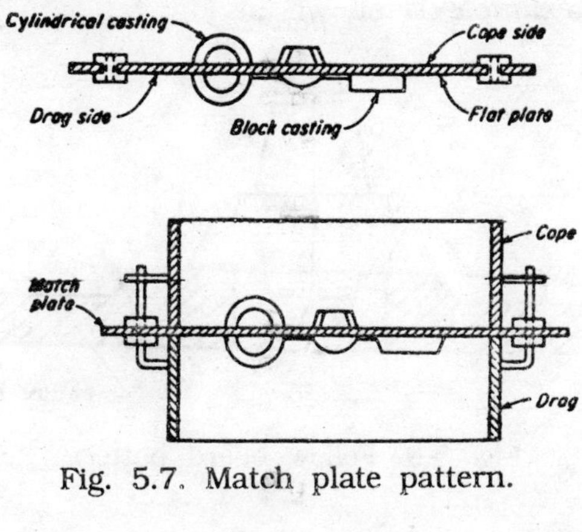

Fig. 5.7. Match plate pattern.

The complete pattern with match plate is entirely made of metal, usually aluminium for its light weight and machinability. But when dimensions are critical, the match plate may be made of steel with necessary case hardening of the critical wear points. Piston rings of the IC engines are produced with the help of match plate patterns.

5. Loose piece pattern : This type of pattern is also used when the contour of the part is such that withdrawing the pattern from the mould is not possible. Hence during moulding the obstructing part of the contour is held as a loose piece by a wire. After moulding is over, first the main pattern is removed and then the loose pieces are recovered the gap generated by the main pattern (Fig. 5.8). Moulding with loose pieces is a highly skilled job and is generally expensive and therefore, should be avoided where possible.

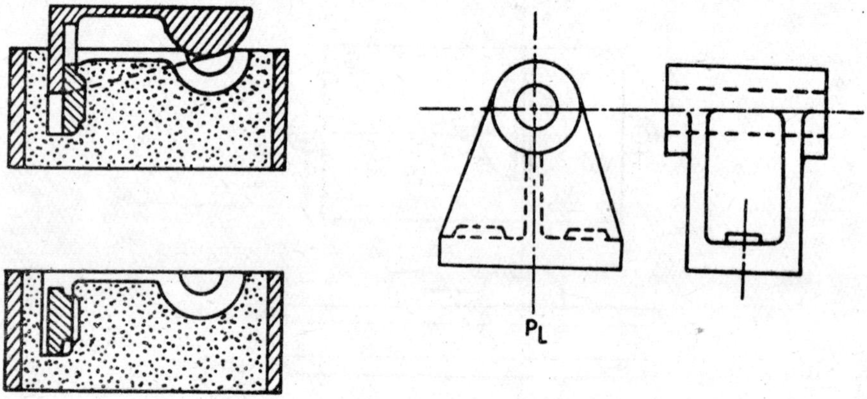

Fig. 5.8. Loose piece pattern.

6. Follow board pattern : This type of pattern is adopted for those castings where there are some portions which are structurally weak and if not supported properly are likely to break under the force of ramming. Hence the bottom board is modified as a follow board to closely fit the contour of the weak pattern and thus support it during the ramming of the drag. During the preparation of the cope, no follow board is necessary because the sand which is compacted in the drag will support the fragile pattern. An example is shown in Fig. 5.9.

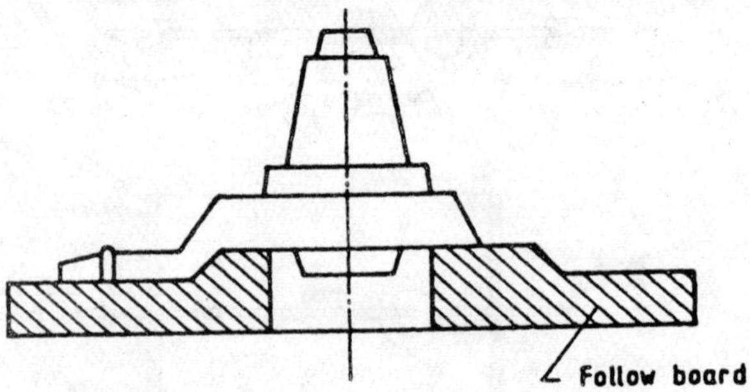

Fig. 5.9. Follow board pattern.

7. Sweep pattern : It is used to sweep the complete casting by means of a plane sweep. These are used for generating large shapes which are axisymmetrical or prismatic in nature such as bell shaped or cylindrical as shown in Fig. 5.10. This greatly reduces the cost of a three dimensional pattern.

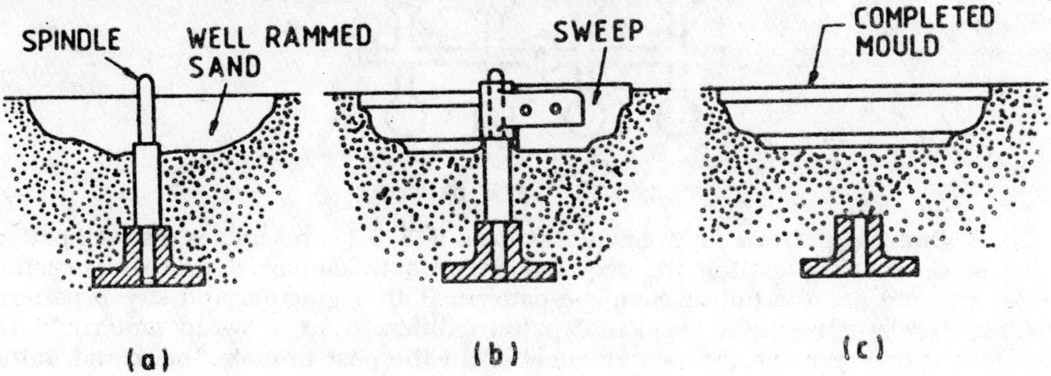

Fig. 5.10. Sweep pattern.

8. Skeleton pattern : A skeleton of the pattern made of strips of wood is used for building the final pattern by packing sand around the skeleton. After packing the sand, the desired form is obtained with the help of a strickle as shown in Fig. 5.11. This type of pattern is useful generally for very large castings required in small quantities where large expenses on complete wooden pattern is not justified.

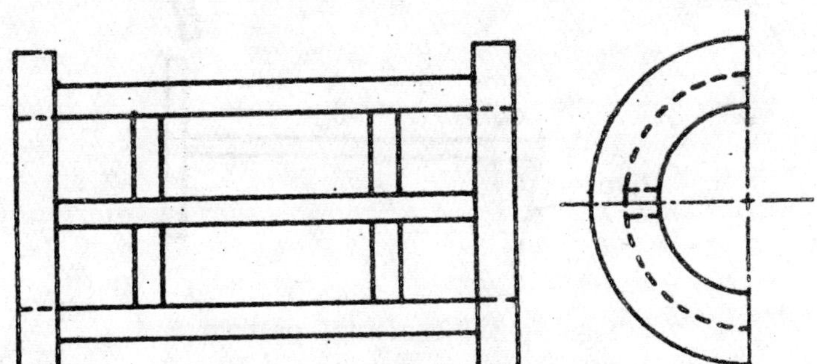

Fig. 5.11. Skelton pattern.

9. Gated pattern : Gated patterns are usually made of metal which increases their strength and reduces the tendency to warp. The sections connecting different patterns serve as runner and gates. This facilitates filling of the mould with molten metal in a better manner and at the same time eliminates the time and labour otherwise consumed in cutting runners and gates. A match when used with a gated pattern saves still more time. A gated pattern (Fig. 5.12) can manufacture many castings at one time and thus it is used in mass production systems. Gated patterns are employed for producing small castings.

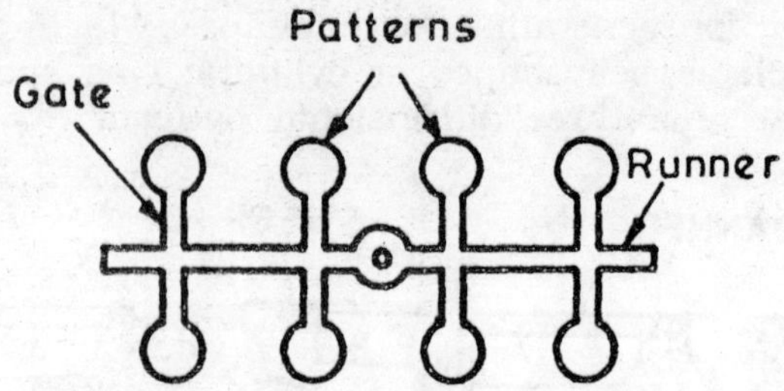

Fig. 5.12. Gated pattern.

10. Segmental pattern : A segmental pattern (Fig. 5.13) resembles a sweep pattern in the sense that for getting the required shape both employ a part or portion of the pattern and not the full or complete pattern. Both segmental and sweep patterns generate circular shapes. A segmental pattern differs from a sweep pattern in the sense that it does not revolve, continuously about the post to make the mould; rather it (segmental pattern) prepares the mould by parts. A segmental pattern completes one portion (stage) of the mould and then moves to the next position to make the next part of the mould and so on till the mould is completed.

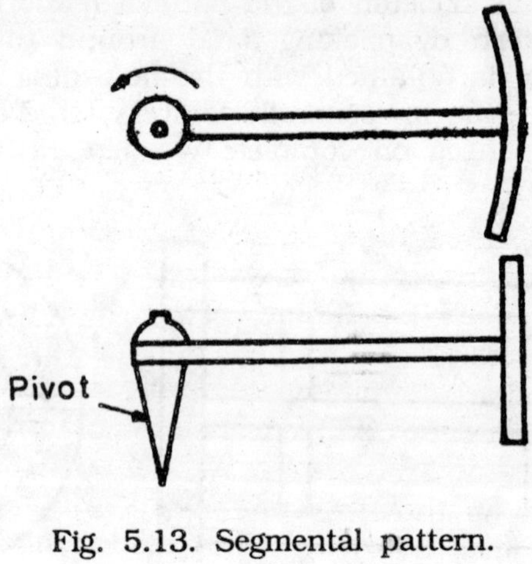

Fig. 5.13. Segmental pattern.

5.4 SAND CASTING

The casting of metal into the intricate shapes that are possible is an immense advantage in engineering construction, and the work of the foundary and pattern shop is of great importance. Castings may be made in steel, cast iron, brass, bronze, aluminium, as well as in numerous other alloys. Sand casting, which is the most common method generally used, will now be given.

5.4.1 Moulding a bracket

Moulding of the bracket of the form shown in Fig. 5.14 (a). The pattern maker will be supplied with a drawing of this as it should be when finished and on the

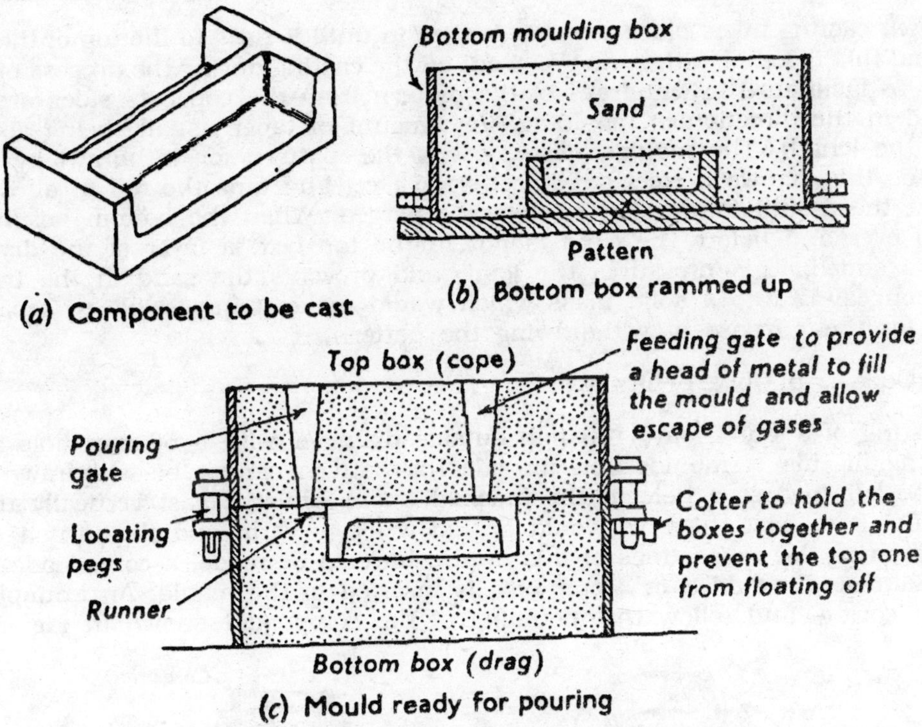

Fig. 5.14 Moulding a bracket.

drawing will be indicated those faces of the bracket which have to be machined. He will make the pattern, leaving contraction allowance, and will leave 3 mm to 5 mm extra on the faces marked for machining. At the foundary the pattern will be used to prepare the mould in moulding boxes, which are steel frames without top or bottom, provided with lugs and holes for lining up when several are put together and with handles for limiting. The pattern will be placed face downwards on a board and a moulding box section laid over it. After a little smooth facing sand mixed with coal dust has been sprinkled on the pattern, the box will be filled with moulding sand and this will be well rammed down with a wooden rammer shaped like a dumb-bell. When the sand has been rammed sufficiently to have made a good impression from the pattern, the top of the box is trimmed off level and it will be like Fig. 5.14 (b). The moulding box is now turned over, another section placed on top of it and rammed with sand as before. A hole is now formed through the sand in the top box for pouring the metal (called the 'pouring gate') and the two boxes are parted again. The pattern is now carefully withdrawn from the bottom box, its impression is touched up with smoothing tools if necessary and the sides dusted with fine plumbago to assist the facing sand in giving a good surface to the casting. Channels (called 'runners') are formed to leaf the metal from the pouring gate to the mould and the top box is replaced. The mould is now ready for pouring and is shown in Fig. 5.14 (c). The top box is called the 'cope' and the bottom one the 'drag'.

When casting takes place metal is poured in until it rises to the top of the pouring gate and this together with runners, is cut off the casting during the process of clearing it up. To facilitate the withdrawl of the pattern from the cope, its sides are slightly tapered in the direction of moulding, the amount of taper usually being about 1 in 50 of the length. The moulder also loosens the, pattern for withdrawl by 'rapping' it. To do this screws a steel rod into its top face, and taps the rod in all directions so that the mould impression is slightly enlarged. When the bottom box has been turned over and before the sand is put in the top box, a layer of the dry parting sand is added. This pressures the joint, and prevents the sand in the two boxes from congealing into a solid mass which would make it impossible to separate the boxes for the purpose of withdrawing the pattern.

5.4.2 Core and Core prints

Casting of a bush, say, 100 mm outside diameter with a 50 mm hole and 150 mm long. If this is moulded on its side the pattern cannot be withdrawn if sand is to be left to cast a hole in the finished bush, whilst if cast vertically the pillar of sand left to cast the hole would be too weak to withstand the inrush of metal when pouring. When castings of this nature have to be made, a core, made of hard, dried sand and moulded in a core-box, is inserted in the mould. An example which include coring, and follow the moulding of the component shown in Fig. 5.15 (a).

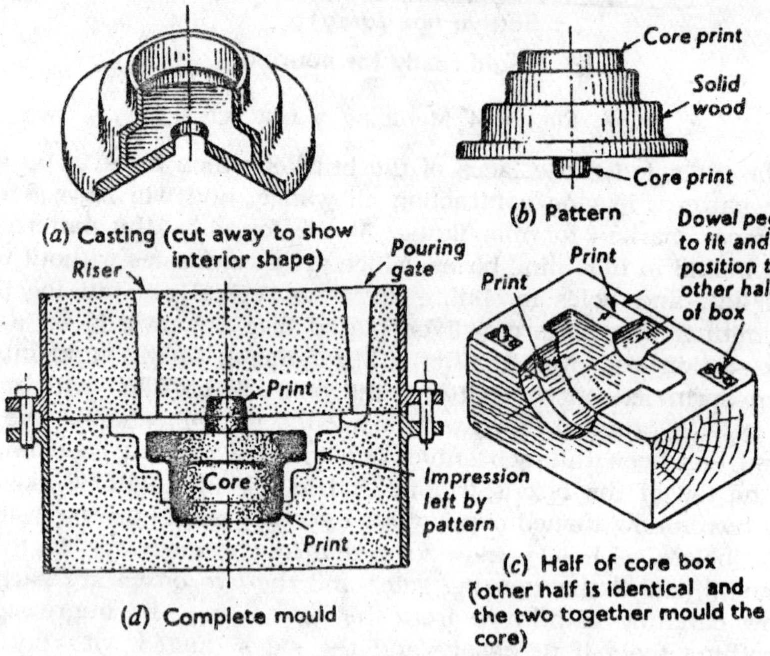

Fig. 5.15 Core and Core prints

The inside of this and the hole in the bottom would be formed by coring and the pattern would be as shown in Fig. 5.15 (b). It will be noticed that the pattern is not only solid where the inside faces of the component should be, but is also provided with projections corresponding with the edges of the metal. These projections are core prints and are for forming recesses in the mould to locate the core. The core for this is made in a separate box called the core box, which is made by the patternmaker, and is auxiliary to the pattern. The core box is often made in two halves to facilitate withdrawl of the moulded core and the one is shown in Fig. 5.15 (c).

At the foundary a core would be moulded in the core box from special core sand and then this core would be dried in an until it was hard. The moulding of the pattern in the cope and drag would leave an impression having a shape similar to the outside of the casting together with the impressions of the core prints. The core itself is made with extremities which just fit these prints when the cope and drag are assembled and when the pattern has been withdrawn the boxes are put together with the core fitted in. This completes the preparation of the mould which will be as shown in Fig. 5.15 (d).

The core-prints serve an important function in keeping the core in its proper position relative to the outside of the mould. If it were not located properly, or moved during pouring, the sides of the casting would not be uniform thickness, and the hole not in proper position.

5.4.3 Split patterns

To facilitate the moulding of some components, the pattern, instead of being made in one piece, is split into two or more parts. These are dowelled so that they may be fitted together accurately to the finished shape. Such an example is given by the flanged pipe shown in Fig. 5.16 (a) and (b) and its moulding would be carried out as follows :

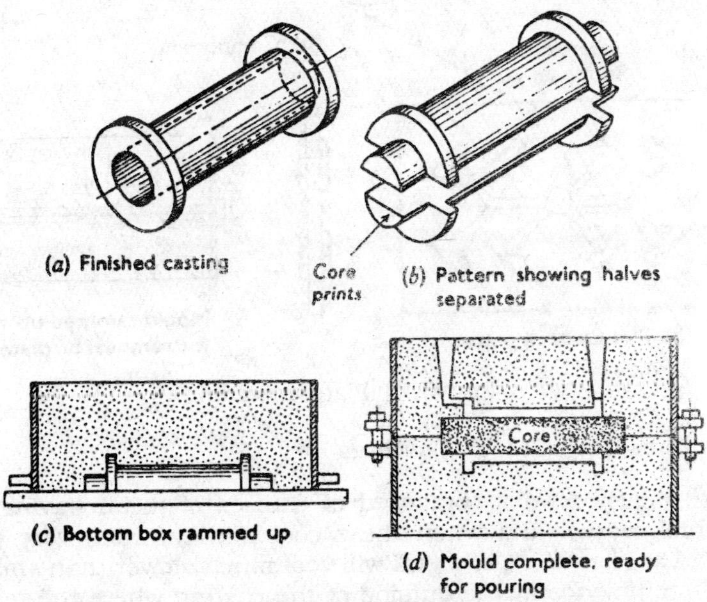

Fig. 5.16 Moulding a flanged pipe.

(1) Ram up one half in bottom box. [Fig. 5.16 (c)]

(2) Turn over, fit the other half and ram top box.

(3) Remove pattern and insert core.

The completed mould ready for pouring is shown in Fig. 5.16 (d).

5.4.4 Plate moulding

When large quantities of fairly simple castings are required they are produced several at once by a process of plate moulding. The form of the part is divided into two halves and the patterns (often in metal) are attached to each side of a board. One side is moulded in the drag, and after turning over, the cope side is moulded. The boxes are then parted and the board and pattern removed. When the boxes are re-assembled without the board, the moulded impressions match up to give the shape of the part and the mould is ready for pouring. Plate moulding is often used in conjunction with moulding machines which vibrate the moulds and do most of the work of ramming. A plate of patterns together with a diagram of the moulding up is shown in Fig. 5.17.

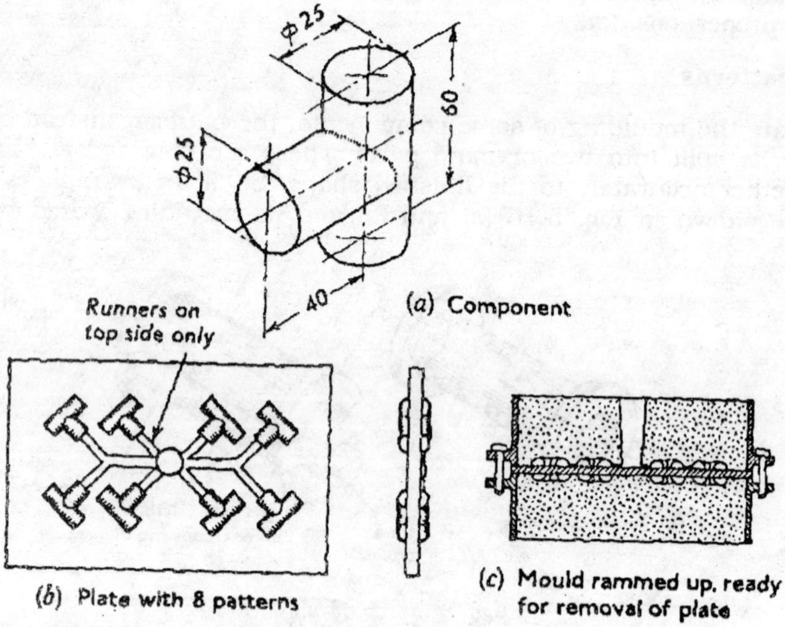

(a) Component

(b) Plate with 8 patterns

(c) Mould rammed up, ready for removal of plate

Fig. 5.17 Plate moulding

5.4.5 Cooling characteristics of Castings

Castings are nearly always composed of masses of metal having various shapes, volumes and thicknesses, and when these cool and contract, very complex stresses are often set up. Large sections of metal will cool much slower than small, thin sections, and cooling will commence at the outside of the casting where the metal is in contact with the cool sand, and will travel inwards.

The effect of cooling on the two parts of a casting are shown in Fig. 5.18 (a) and (c). At Fig. 5.18 (a) the sides of the L will commence to cool first and cooling will proceed inwards. As cast iron cools the crystals tend to set themselves in the direction of cooling. This is shown shaded in Fig. 5.18 (a) and results in a plane of weakness at the corner. For this reason, and to help are well rounded off by fillets as shown in Fig. 5.18 (b). At Fig. 5.18 (c) a large boss of metal is joined by arms to a much thinner section. When cooling from the molten state, the thin sections will have solidified and be well cooled down before the large volume of metal at the boss has cooled very much. Ultimately this will cool down and contract, but as the arms have now cooled and are rigid, the force exerted by the contraction of the large boss cannot be relieved by any movement on the part of the two thin sections. This means that a permanent internal stress is left in the casting between the large boss and the outer frame, and even if the casting operated to be all right, a sudden blow or shock might cause it to fracture at one of the sections shown in Fig. 5.18. Machining the hard and rigid skin from the casting on its outer face would relieve some of the stress and probably allow a certain amount of spring to take place.

These two simple cases will serve to show how internal stresses and weaknesses may exist in a casting and although these effects by studying the method of casting, they cannot be prevented altogether. Castings with internal stresses will often warp when these stresses have been relieved by machining off the hard and rigid outer skin. When it is important that the shape of a casting shall remain stable, e.g. for a gauge, large worm wheel, etc., it should be rough machined and 'weathered' or 'seasoned' by being placed in the open for a few months. This allows a dissipation of the internal stresses to take place and minimizes the possibility of distortion after the casting has been finished. For small castings a quick method of seasoning is to place a number of castings in a tumbling barrel for about half an our. The vibration and light blows which the castings relieve whilst being tumbled round in the barrel form a very effective method of relieving internal stresses.

EXERCISES

A. Theory

5.1. Define foundry.

5.2. Draw a neat sketch showing the sand mould making procedure.

5.3. Difference between pattern and casting.

5.4. List the functions of a pattern.

5.5. Write the advantages, limitations and the field of applications of pattern.

5.6. Write the advantages, limitations and application of wood material.

5.7. Name the different pattern materials and explain any one.

5.8. Write the pattern allowances.

5.9. List the various types of pattern types and explain with the help of sketches.

5.10. Name the different foundry tools and equipment. Explain the hand tools with the help of neat sketches.

5.11. Name the various types of moulding boxes used in foundry.

5.12. Name the various type of ladles used in foundry.

5.13. Name the various types of moulding materials used in foundry and explain them clearly.

5.14. Differentiate between temporary moulds and permanent moulds.

5.15. Write down the different processing properties of moulding materials.

6

WELDING, BRAZING AND SOLDERING

6.1. INTRODUCTION

The soldering process requires temperatures below 450°C and often below 300°C. The brazing is a process similar to that of soldering but involves temperatures above 450°C. The term 'braze' comes from the use of brass as the substance used to make the joint. The procedure for brazing is similar to that for soldering, but the end result is a stronger joint than that given with solder.

With brazing and soldering, the joint is effected by inserting a metal between the two metal surfaces being joined, the inserted metal having a lower melting point than that of the materials being joined with welding, the joint is effected directly between the parts being joined by the application of heat or pressure.

Welding processes are capable of producing high strength joints. The temperature involved in producing the welds may, however, cause determental changes in the materials being joined. These may be local distortions due to uneven thermal expansions, residual stresses, or micro structural changes.

Fusion welding is a major category of welding and consists of consumable or unconsumable electrode arc welding, and high energy- beam welding processes. Fusion welding is not used in a modern industrial setting and are used basically for maintenance and repair work.

Solid-state welding, joining takes place without fusion, thus, there is no liquid (molten) phase in the joint.

6.2. WELDING

Welding is a process of joining similar metals by application of heat with or without application of pressure and addition of filler material.

Fusion welding - In fusion welding an external heat source is used to melt the interfaces of the joint materials and so cause the materials to fuse together.

Solid-state welding - Pressure is used to bring the two interfaces of the joint materials into intimate contact and so fuse the two materials together.

6.3. GAS WELDING

It is a non pressure fusion welding process and includes all processes in which gas is used as a source of heat to melt the ends of the pieces to be joined on solidification. A filler metal is needed in welding of sheets above 1.5 mm thickness but no filler metal is needed for welding below 1.5 mm thickness.

The oxy-acetylene flame is most widely used as it produces very high temperatures (3500°C) and can be used for welding of a variety of ferrous and non ferrous materials.

6.4. OXY-ACETYLENE GAS WELDING

In oxy-acetylene gas welding, the heat source is a flame produced by burning a mixture of oxygen and acetylene gases. The temperature of the flame is about 3500°C which is more than sufficient to melt mostly commonly used metals. Since the gases are expensive this process is used only for sheet metal and thin plate. Fig. 6.1 shows a basic set of gas welding equipment.

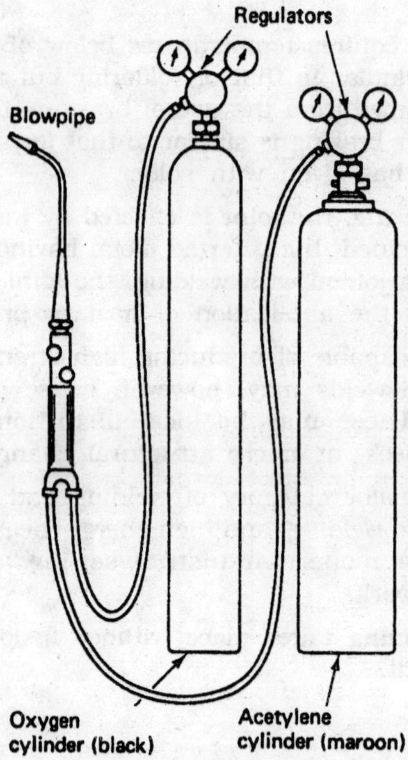

Fig. 6.1. Basic gas welding equipment.

The size of the blow pipe nozzle can be changed to allow variations in the volume of gas burnt and, correspondingly, the heat output of the blow pipe. This allows the heat energy output to be matched to the thickness and composition of the material being joined.

6.4.1 Gas Cylinders

As shown in Fig. 6.2, during welding, the oxygen and acetylene and ordinarily obtained from strong steel cylindrical containers, which are filled by commercial suppliers of these gases. The oxygen cylinders are usually filled at a pressure of 12.5 to 14 MPa and have capacity of 6.23 m^3. The mass of oxygen cylinder is approximately 66 kg when its full. The standard colour for the oxygen cylinder is black. The acetylene is stored in cylinders under an initial pressure of 15 atmospheres and have the capacity of 7.6 m^3. It mass is about 86kg when it is full. The acetylene cylinders are equipped with fusible plugs at the bottom in order to save the cylinder from explosion in case of fire. The standard colour for the acetylene cylinder is maroon.

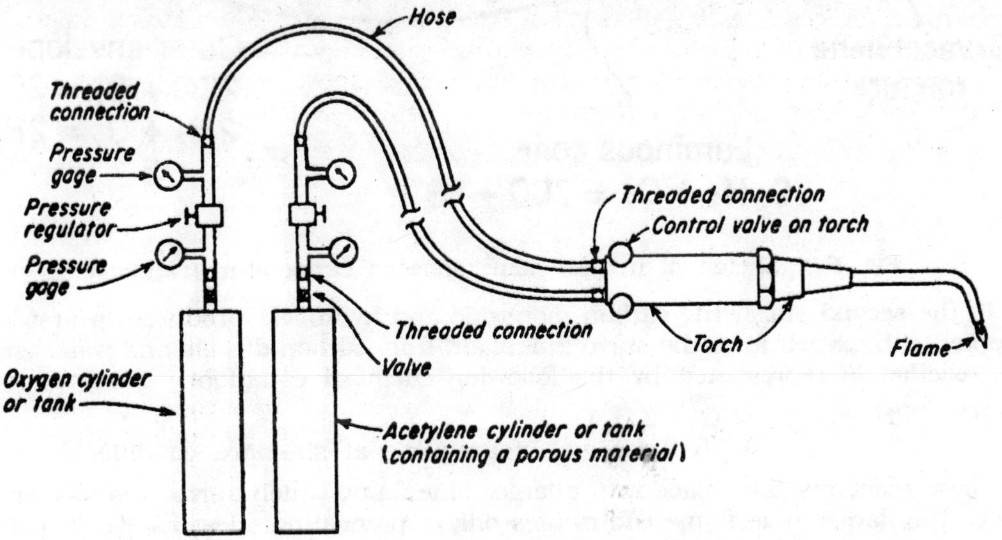

Fig. 6.2 Schematic of the equipment used for oxyacetylene welding.

The pure acetylene is not stable at pressures above atmospheric pressure and may decompose explosively. Therefore, in order to store acetylene at 15 atmospheres, it is dissolved in acetone which is capable of absorbing a large volume of acetylene and of releasing it as the pressure falls. One volume of acetone at 15°C and at atmospheric pressure dissolves about 25 volumes of acetylene. The dissolving capacity grows in proportion to the pressure so that at the same temperature but at 15 atmospheres acetylene pressure, one volume of acetone absorbs 25 x 15 = 375 volumes of acetylene. The cylinder is first filled with a dry porous material (e.g. charcoal or asbestos) which keeps the acetylene well separated. Then the tank is partially filled with acetone and acetylene is dissolved in it under a pressure of 15 atmospheres.

6.4.2 Gas flames

The combustion of acetylene with pure oxygen at the tip of a gas welding torch takes place in two stages. In the first stage, the carbon from the acetylene combines with oxygen to form carbon monoxide, the hydrogen of the acetylene being freed. The following chemical reaction occurs.

$C_2H_2 + O_2 = 2CO + H_2$ + (448 kJ/mole of acetylene at standard conditions)

This reaction takes place within the inner, brilliant white cone flame, close to the tip opening, as shown in Fig. 6.3. This provides the most concentrated heat with the highest temperature for welding.

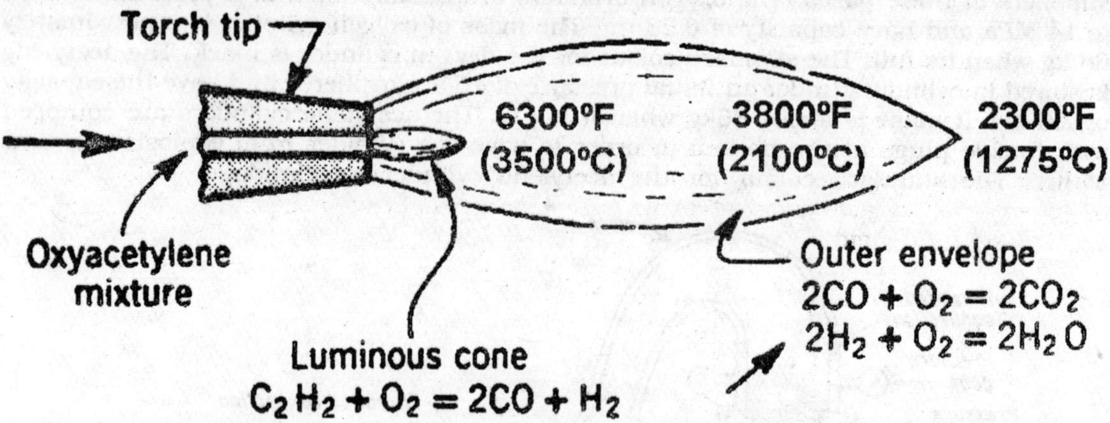

Fig. 6.3. Sketch of neutral flame showing temperatures attained.

In the second stage, the carbon monoxide and hydrogen (produced in first stage), together with oxygen from the surrounding air, from carbon dioxide and water (steam). The reaction is represented by the following chemical equation.

$$4CO + 2H_2 + 3O_2 = 4CO_2 + 2\,H_2O$$

(+ 812 kJ/mol of acetylene at standard conditions)

These reactions take place in the larger blue flame which surrounds whitish cone flame. This larger blue flame contributes only a preheating effect for welding. It also protects the molten metal from oxidation.

There are three types of flames obtainable for gas welding.

(1) Neutral flame : A neutral flame is obtained by supplying equal volumes of oxygen and acetylene. The neutral flame, as shown in Fig. 6.4(a), has the inner cone is a blunt 'U'. The most of the oxy-acetylene welding is done with the use of neutral flame. (e.g. steel, cast iron, copper, aluminium etc.).

(2) Oxidising flame : This flame is obtained when there is an excess of oxygen present. It is a roaring, noisy flame with a sharp, conical, brilliant white inner cone. It is similar to neutral flame but the inner cone is less luminous and shorter, as shown in Fig. 6.4 (b).

The oxidising flame cannot be used in most applications of oxyacetylene welding, since the metal is badly oxidized. However, it may preferable in rare cases. It is used for copper and aluminium alloys.

(3) Carburising or reducing flame : Carburising or reducing, flame is obtained by supplying an excess of acetylene. This flame, as shown in Fig. 6.4(c), Since this

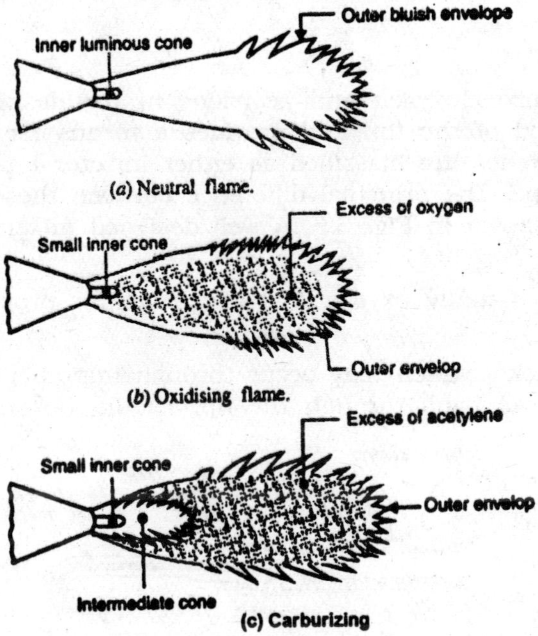

Fig. 6.4 (i). Types of flames

flame diffuses carbon into the surf- aces of low carbon steels it can be used for flame hardening processes, the flame being moved slowly over the metal surface followed by a quenching spray. It is also used for depository hard facing materials such as 'stellite' on to steel component. It has three zones.

(i) An inner zone, (ii) An intermediate cone of whitish colour, (iii) An outer cone of bluish colour.

Fig. 6.4 (ii) shows the photographs of flames.

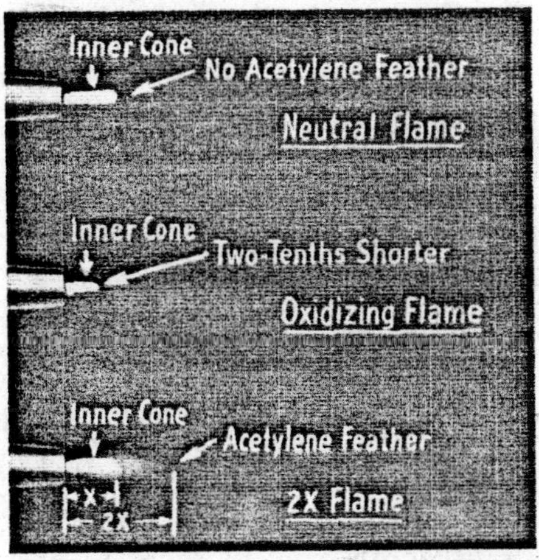

Fig. 6.4(ii) Photograph of neutral, oxidizing and carburizing flame characteristics.

6.4.3 Welding torch

A welding torch mixes oxygen and acetylene in the desired proportions, burns the mixture at the end of the tip, and provides a means for moving and directing the flame. Welding torches are classified as either injector type or positive pressure (or equal pressure) type. The principal difference between these two types of torches is in the mixers, as shown in Fig. 6.5. A well designed mixer should be able to do these two things :

(1) Mix the gases, usually oxygen and acetylene, for proper combustion at the tip.

(2) Arrest flash backs, which may occur through improper operation. Any flame, which happens to travel back through the tip, should be stopped at the mixer.

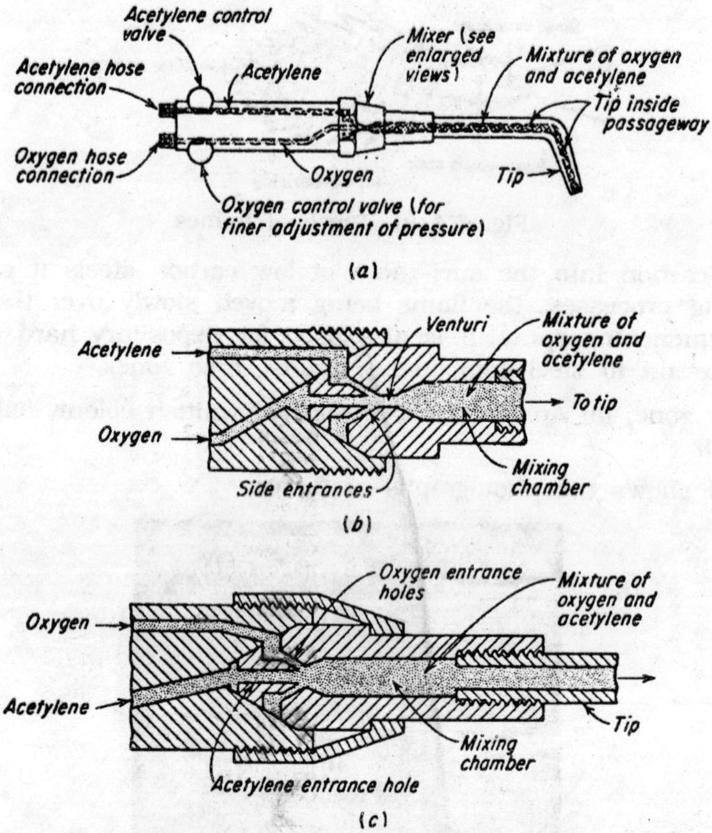

Fig. 6.5 Oxyacetylene welding torch. (a) Oxyacetylene welding torch (b) An injector type mixer (c) Equal pressure or positive pressure type mixer.

Welding torch tips are made of a high thermal conductivity material, such as copper. This helps to keep the tip sufficiently cool in order to prevent melting of the tip and any flash backs due to overheating of the tip.

6.4.4 Oxyacetylene Welding technique

Fig. 6.6 shows a weld zone of oxyacetylene welding.

When igniting a torch, the acetylene is first turned on, using the control valve on the torch, and then it is ignited with a friction spark lighter. Following this, the oxygen is turned on, and the flame is adjusted. The weld is started by preheating and melting a small puddle of molten base metal. A welding rod, if used to provide filler metal, should be held in the flame so that its end melts at about the same time as the base metal. The speed of welding should be as uniform as possible, and it is determined by the speed of melting. To obtain proper penetration and to produce a good looking weld, the tip may be moved with side-to- side motion, slowly and uniformly, following a curved path along the forward edge of the molten metal puddle. While doing this, the tip is advanced uniformly and slowly along the desired weld.

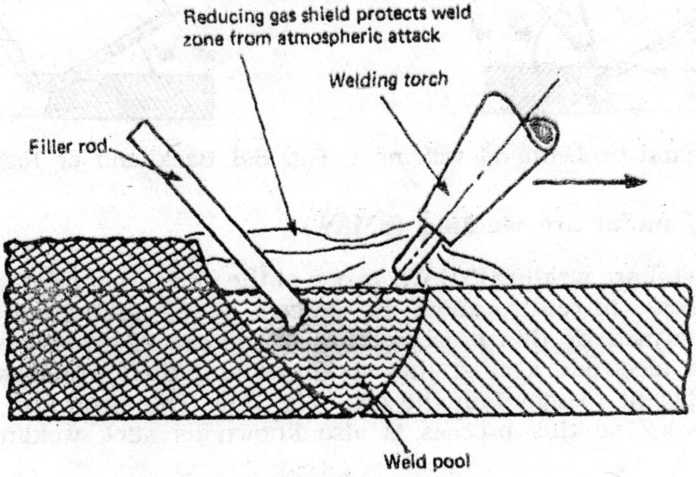

Fig. 6.6. Oxy-acetylene welding

(a) Forehand or Lefthand welding : In forehand welding (Fig. 6.7) the flame is directed forward in the direction toward which the weld is being made. It is usually recommended for welding thinner pieces (below 1/8 inch thick).

(b) Backhand or Right-hand welding : In backhand welding (Fig. 6.8) the pressure of the flame helps to control the molten metal puddle, especially in positions such as overhead or vertical, where the molten metal may easily flow out of a weld. Backhand welding is usually recommended for welding thicker pieces, is accomplished when the flame is directed backward.

6.5. ARC-WELDING

In arc welding, the heat required is obtained through electrical energy. Using either a consumable or unconsumable electrode (rod or wire), an arc is produced between the tip of the electrode and the work piece to be welded, using AC or DC power supplies. This arc produces temperatures in the range of 30000°C, which is much higher than those developed in oxyfuel gas welding.

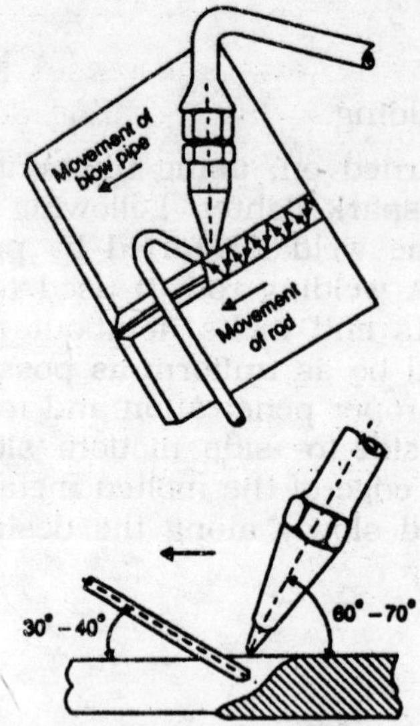

Fig. 6.7 Forehand or Lefthand welding

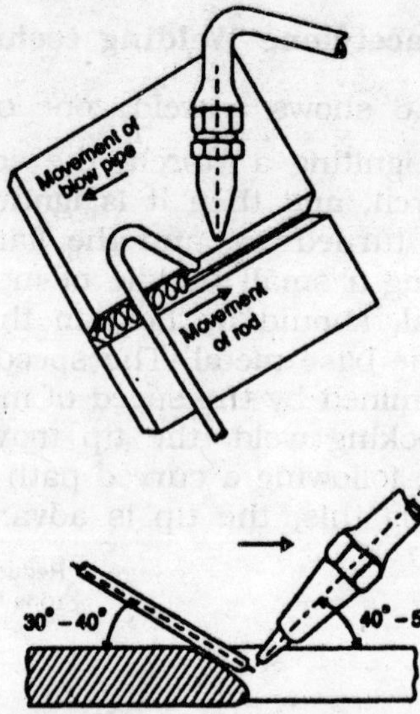

Fig. 6.8 Backhand or Right-hand welding

6.5.1. Shielded metal arc welding (SMAW)

Shielded metal arc welding (SMAW) is one of the oldest, simplest, and most versatile joining processes. Currently, about 50 percent of all industrial and maintenance welding is performed by this process. The electric arc is generated by touching the tip of a coated electrode against the workpiece and then withdrawing it quickly to a distance sufficient to maintain the arc. (Fig. 6.9). The electrodes are in the shape of thin, long stick so this process is also known as stick welding.

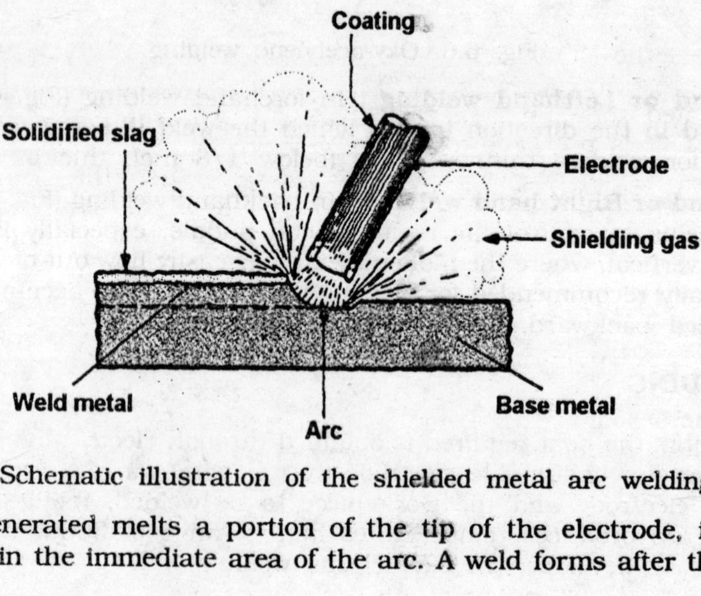

Fig. 6.9. Schematic illustration of the shielded metal arc welding process.

The heat generated melts a portion of the tip of the electrode, its coating, and the base metal in the immediate area of the arc. A weld forms after the molten metal a mixture of the base metal (workpiece), electrode metal, and substances from the coating on the electrode- solidifies in the weld area. The electrode coating deoxidizes and provides a shielding gas in the weld area to protect it from oxygen in the environment.

A bar section at the end of the electrode is clamped to one terminal of the power source, while the other terminal is connected to the workpiece being welded (Fig. 6.10). The current usually ranges between 50A and 300A, with power requirements generally less than 10kW. Too low a current causes incomplete fusion, and too high a current can damage the electrode coating and reduce its effectiveness. The current may be DC or AC. For sheet-metal welding, DC is preferred because of the steady arc.

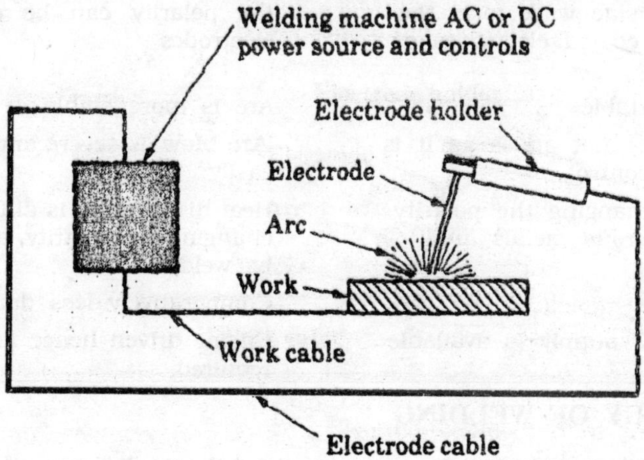

Fig. 6.10. Schematic illustration of the shielded metal arc welding operation.

The **polarity** of the DC current, i.e., direction of current flow, can be important, depending on factors such as type of electrode, the metals to be welded, and the depth of the heated zone. In **straight polarity,** the workpiece is positive and the electrode negative, it is preferred for sheet metals because of shallow to penetration and for joints with very wide gaps. In **reverse polarity,** where the electrode is positive, deeper weld penetration is possible. In the AC method, the arc pulsates rapidly; it is suitable for welding thick sections and using large diameter electrodes at maximum currents.

This process is commonly used in general construction, ship building, and pipelines, as well as for maintenance work, since the equipment is portable and can be easily maintained. It is especially useful for work in remote areas where portable fuel-powered generators can be used as the power supply. The SMAW process is best suited for workpiece thicknesses of 3 to 19 mm, although this range can be easily extended using multiple-pass techniques.

The SMAW process has the advantage of being relatively simple and versatile, requiring a relatively small variety of electrodes. The equipment consists of a power supply, power cables and electrode holder.

6.5.2 Comparison between A.C. and D.C. arc welding

Following is the comparison between AC and DC arc welding is given in the Table 6.1

Table 6.1 Comparison between AC and DC arc welding

SN	AC arc welding	DC arc welding
1	Transformer cost low and its maintenance cost is low	Generator costs more and its maintenance cost is high
2	Transformer has no moving parts and is simpler	Welding generator has rotating parts and is more complicated.
3	All types of electrodes cannot be used because sine wave form the current. Only coated electrodes can be used.	All types of electrodes can be used because the polarity can be changed to suit the electrodes
4	Arc is never stable	Arc is more stable
5	'**Arc blow**' does not arises as it is very easy to control	'**Arc blow**' is severe and cannot be controlled easily.
6	No need for changing the polarity. Hence only ferrous metals are usually welded.	Heat distribution is different in two poles. By changing the polarity, all types of metals can be welded.
7	More dangerous	Comparatively less dangerous
8	Used when AC supply is available	Engine driven hence no AC mains supply is required.

6.6. METALLURGY OF WELDING

Every welding process involves a complete thermal cycle of heating. The metal is heated over a range of temperature (up to fusion) followed by cooling to ambient temperatures. In welding, differential heating of the workpiece takes place, the temperature peaks at the weld area. It result in a complex mixture of micro-structures particularly in steel. Welding also produces internal stresses and plastic strain in the workpiece. If the welding atmosphere is not properly controlled certain chemical changes may also take place. It shows that irrespective of the process involved, thermal, chemical and structural changes take place during welding.

In all welding processes the flow of heat in the weld zone is highly directional towards the adjacenet cold metal. It produces columnar grains. Thus the original microstructure of steel containing ferrite and pearlite is changed into another microstructure. The composition of the weld at the beginning and end is also different. The molten metal has also good capacity of the weld to absorb gases also decreases. Thus absorbed gases escapes from the metal creating prosity or gas pockets in the weld.

Practically every type of steel can be welded by one of the welding processes. Steels contain various alloying elements like carbon, silicon, manganese, nickel, chromium, molybdenum, vanadium, aluminium, titanium, and copper. These alloying elements exert their influence on the microstructure of steel in one or more of the following ways during welding across the welding joint.

(i) By forming a solid solution.

(ii) By forming carbides.

(iii) By forming intermediate compounds.

(iv) By altering the strength, toughness and hardness.

(v) By altering or controlling the grain size.

(vi) By altering the transition temperature. Steels are generally annealed, normalised or stress relieved to get almost homogeneous microconstituents in welds.

The welding of various metals is listed in Table 6.2.

Table 6.2 Welding of various metals.

S.N.	Metal	Welding process	Remarks
1	Wrought iron	All commonly used welding process	Slag acts as flux and protects the metal
2	Ingot iron	—do—	Excellent welding characteristics.
3	Mild steel	All welding processes.	Good welding characteristics.
4	Medium carbon steels	Various fusion processes.	In some cases preheating and heat treatment is required.
5	High carbon steels	Gas, shielded metal arc, submerged arc, thermal welding and resistance welding	Difficult to weld due to its high carbon content.
6	Tool steels	—do—	Quite difficult to weld, pre- heating is required
7	Grey cast iron	Gas welding, shielded metal arc welding, thermit welding	Preheat the component above 850°C. After welding cool very slowly in suitable insulating material.
8	Malleable cast iron	—do—	—do—
9	Low alloy steels	Gas, shielded arc, inert, gas, submerged arc, atomic hydrogen and resistance welding	Preheating and postheating of workpiece is essential.
10	Chromium steels	Gas welding, resistance welding, pressure gas welding, atomic hydrogen welding, inert gas welding.	—do—
11	Stainless/heat resisting steels	Atomic hydrogen inert gas, shielded arc, resistance welding and gas welding.	Heat treatment after welding processes is essential.
12	Copper and its alloys	Commonly used welding methods.	Reducing flux is needed.
13	Aluminium and its alloys	Gas, electric resistance, shielded arc, soldering and brazing process, TIG and MIG.	Heat treatment after welding is necessary.
14	Magnesium and its alloys	TIG, MIG, Oxy-acetylene, and resistance welding.	Reducing flux is needed.

6.7 ELECTRODES

Electrodes for the consumable arc welding processes are classified according to the strength of the deposited weld metal, current (AC or ± DC) and the type of coating. Electrodes are identified by numbers and letters or by colour code, particularly if they are too small to imprint with identification. Typical coated electrode dimensions are 150-160 mm long and 1.5-8 mm in diameter. The thinner the sections to be welded and the lower the current required, the smaller the diameter of the electrode should be.

The specifications for the electrodes and filler metals, including tolerances, quality control procedures, and processes, are stated by AWS and ANSI and AMS by the Society of Automotive Engineers (SAE). Among others, the specification require that the wire diameter not vary more than 0.05 mm from nominal size, and that the coatings be concentric with the wire. Electrodes are sold by weight and are available in a wide variety of sizes and specifications. Selection and recommendations for electrodes for a particular metal and its application can be found in supplier literature and various handbooks.

Electrode coatings

Electrodes are coated with claylike materials that include silicate binders and powdered materials such as oxides, carbonates, fluorides, metal alloys and cellulose (Cotton cellulose and wood flour). The coating which is brittle and has complex interactions during welding, has the following basic functions :

1. Stablize the arc.
2. Generate gases to act as a shield against the surrounding atmosphere. The gases produced are carbon dioxide and water vapour, and carbon monoxide and hydrogen in small amounts.
3. Control the rate at which the electrode melts.
4. Act as a flux to protect the weld against formation of oxides, nitrides, and other inclusions, and with the resulting slag, protect the molten weld pool.
5. Add alloying elements to the weld zone to enhance the properties of the weld, including deoxidizers to prevent the weld from becoming brittle.

The deposited electrode coating or slag must be removed after each pass in order to ensure good weld. A manual or power wire brush can be used for this purpose. Base electrodes and wire, made of stainless steels and alluminium alloys are also available. They are used as filler metals in various welding operations.

6.8. ADVANTAGES AND LIMITATIONS OF THE OXY-ACETYLENE WELDING PROCESSES

Advantages

1. Low capital cost.
2. Portability (self contained energy source)

3. Versatility : hard soldering, brazing, braze-welding and hard facing in all metals as well as being suitable for fusion welding in all positions.
4. Leftward welding : mild steel 0.8 mm thick at 6-8 m/h, to 5 mm thick at 4-5 m/h.
5. Rightward welding : mild steel 5 mm thick 4-5 m/h, to 25 mm thick at 1-2 m/h.
6. Given good instruction it is relatively easy process in which to acquire skills.

Limitations

1. Compressed gases are very dangerous and extreme care is required in their handling and storage.
2. Cylinder rental is expensive
3. Cost of the gas is expensive particularly which welding thick material.

6.9. ADVANTAGES AND LIMITATIONS OF THE MANUAL METALLIC ARC WELDING

Advantages

1. Since the electrical supply is normally taken from the mains, the energy available is virtually unlimited.
2. The process is quicker than for gas welding as the heat output of the arc is greater than that of a gas flame.
3. The cost of the weld per metre run is usually lower than for gas welding due to the speed of the process and the cheaper energy source.
4. Multi-run welds are normally used for metal above 10 mm thick to limit the welding current and avoid 'splatter'.
5. Due to the more rapid heating of the parent metal, the heat affected zone is more localised, resulting in more rapid cooling, less grain growth, and a stronger weld than with gas welding.
6. The hazards associated with arc welding are fewer than those associated with gas welding, providing correct protective clothing is worn and the electrical conductors and connections are maintained in good conditions.

Limitations

1. Higher initial cost than for gas welding equipment.
2. Less portable than gas welding a mains supply of electricity required, or a large diesel generating set for site work.
3. Coated electrodes are very much more expensive than gas welding filler rods.
4. The arc is difficult to control when welding thin material.
5. The process has fewer applications than for gas welding.
6. Considerable skill is required to make a good quality weld.

6.10. BRAZING AND SOLDERING

This section describes two joining processes that involve lower temperatures than those required for welding. In both brazing and soldering, filler metals are placed in or supplied to the joint. They are then melted using an external source of heat and, upon solidification, a solid joint is produced. Brazing temperature are higher than those for soldering, and the brazed joint has higher strength. Thus brazing and soldering are arbitrarily distinguished by temperature.

6.11. BRAZING

Brazing is a joining process in which a filler metal is placed at or between the faying surfaces to be joined, and the temperature is raised to melt the filler metal but not the workpieces (Fig. 6.11a). The molten metal fills the closely fitting space by capillary action. Upon cooling and solidification of the filler metal, a strong joint is obtained. Actually, there are two types of brazing processes : (1) that which we have already described and (2) braze welding (Fig. 6.11b), in which the filler metal is deposited at the joint with a technique similar to oxyfuel gas welding. A diagram of brazing blowpipie is shown in Fig. 6.12 and brazing hearth, with motor driven rotary blower, is shown in Fig. 6.13.

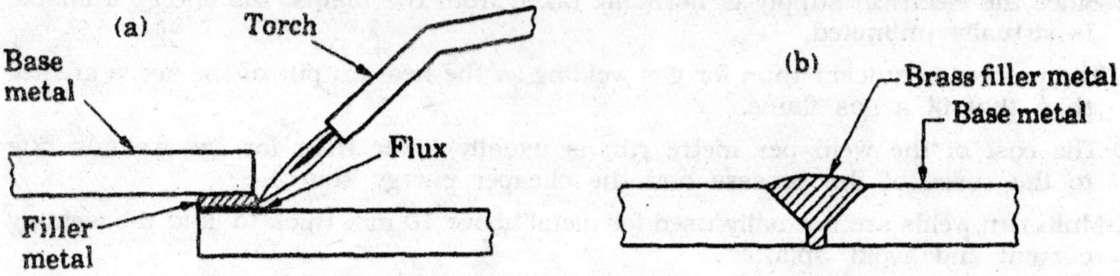

Fig. 6.11 (a) Brazing and (b) braze welding operations.

Fig. 6.12. Blow pipe.

Filler metals used for brazing melt above 450°C. The temperatures employed in brazing are below the melting point (solidus temperature) of the metals to be joined. Thus this process is unlike liquid state welding processes in which the workpieces must melt in the weld area for fusion to occur. Problems associated with heat - affected zones, warping, and residual stresses are therefore reduced in brazing.

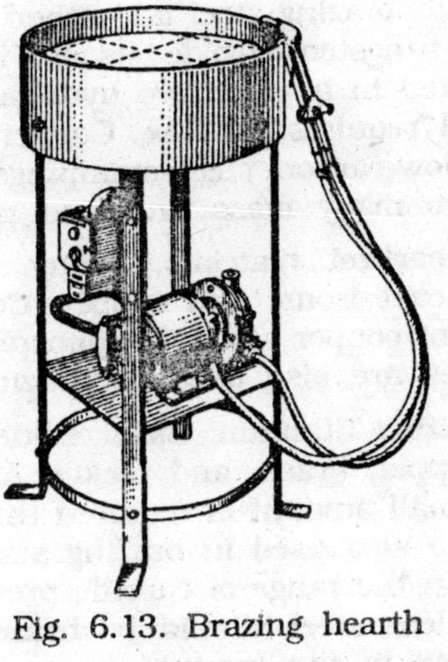

Fig. 6.13. Brazing hearth

The strength of the brazed joint depends on (1) joint design and (2) the adhesion at the interfaces of the workpiece and filler metal. Consequently, the surface to be brazed should be chemically or mechanically cleaned to ensure full capillary action; hence the use of flux is important.

6.11.1. Brazing metals

The most commonly used brazing metals are copper and copper alloys, silver and silver alloys, and aluminum alloys. Table 6.3 lists some of the frequently used brazing metals, compatible processes, and the materials that are commonly brazed.

Table 6.3. Commonly used Brazing Metals and their Uses.

Braze Metal	Composition	Brazing process	Base Metals
Brazing brass	60% Cu, 40% Zn	Torch Furnace Dip Flow	Steel, copper, high-copper alloys, nickel, nickel alloys, stainless steel.
Manganese bronze	58% Cu, 1% Sn 1% Fe, 0.25% Mn 39.5% Zn	Torch	Steel, copper, high-copper alloys, nickel, nickel alloys, stainless steel
Nickel silver	18% Ni, 55-65% Cu, 27-17% Zn	Torch, Induction	Steel, nickel, nickel alloys
Copper silicon	1.5% Si, 0.25% Mn 98.25% Cu, 1.5% Si, 1.00% Zn, 97.5% Cu	Torch	Steel
Silver alloys (no phosphorus)	5-80% Ag, 15-22% Cu, balance Zn + Sn + Cd	Torch, Furnace, Induction, Resistance	Steel, copper, copper alloys, nickel, nickel alloys, stainless steel
Silver alloys (with phosphorus)	15% Ag, 5% P, 80%Cu	Torch, Furnace, Induction, Resistance Dip	Copper, copper alloys
Copper phosphorus	93% Cu, 7% P	Torch, Furnace, Induction Resistance	Copper, copper alloys

Copper is used primarily for brazing steel and other high-melting point materials, such as high-speed steel and tungsten carbide. Its use is confined almost exclusively to furnace operations conducted in a protective hydrogen atmosphere, in which the copper is extremely fluid and requires no flux. Copper brazing is used extensively for assemblies composed of low-carbon steel stampings, screw-machine parts, and tubing - materials common to many mass produced products.

Copper alloys were the earliest mateials. Today, copper-zinc allys are used extensively for brazing steel, cast irons, and copper. Copper phosphorus alloys are used for the fluxless brazing of copper since the phosphorus can reduce the copper oxide film. Manganese bronzes are also used as brazing material.

Pure silver is used in brazing titanium. Silver solders, silver-copper-zinc alloys, are used in joining steels, copper, brass, and nickel. Although these brazing alloys are quite expensive, such a small amount is required that the cost per joint is really quite low. The silver alloys are also used in brazing stainless steels. However, since the brazing temperatures are in the range of carbide precipitation (sensitization), only stabilized or low-carbon stainless steel should be brazed with these alloys if good corrosion resistance is required in the product.

Aluminum-silicon alloys, containing about 6% to 12% silicon, are used for brazing aluminum and aluminum alloys. By using a braze metal that is not greatly unlike the base metal, the possibility of galvanic corrosion is reduced. However, since these brazing alloys have melting points of about 1113°F (610°C), and the melting temperature of commonly brazed aluminum alloys such as 3003 is around 1290°F (670°C), the control of the brazing temperature is quite critical. In brazing aluminum, proper fluxing action, surface cleaning, and/or the use of a controlled-atmosphere or vacuum environment is required to assure adequate flow of the braze metal.

A commonly used procedure when brazing aluminum is to use sheets of metal that have one or both surfaces precoated with the brazing alloy to a thickness of about 10% of the total sheet. Bonds are made by simply coating the joint area with a suitable flux, placing the materials in contact, and heating. The brazing sheets have sufficient coating to form adequate fillets.

Amorphous alloy brazing sheets have been produced by cooling metal at rates in excess of 1 million degrees Centigrade per second. The resulting metal foils are extremely thin, 0.0015 in. (0.04 mm) being typical, and exhibit excellent ductility and flexibility in the amorphous condition (even with alloys that are characteristically brittle in their crystalline condition). Shaped inserts can be cut or stamped from the foil, inserted into the joint, and heated. The resulting braze exhibits uniform melting and penetration. Since the braze material is fully dense when inserted, no shrinkage or movement is observed during the brazing operation. A variety of brazing alloys are currently available in the form of amorphous foils.

A nickel-chromium-iron-boron brazing alloy is frequently used to braze heat-resistant alloys that are to be used at high temperatures. Althoguh the service termperature of the product will be above the melting point of the braze alloy, the braze material will not melt. During the brazing operation, the boron in the brazing alloy diffuses into the base metal, raising the melting point of the remaining alloy to above the subsequent service temperature.

Cast iron is not readily wettable because of the graphite. Consequently, before cast iron can be brazed, the graphite must be removed by etching.

6.11.2. Filler metals

Several filler metals (braze metals) are available and have a range of brazing temperatures (Table 6.4). They come in a variety of shapes, such as wire, rings, shims, and fillings. The choice of filler metal and its composition are important in order to avoid embrittlement of the joint (by grain boundary penetration of liquid metal), formation of brittle intermetallic compounds at the joint, and galvanic corrosion in the joints. Note that filler metals for brazing unlike other welding operations, generally have significantly different compositions than the metals to be joined.

Table 6.4. Typical filler metals for brazing various metals and alloys

Base metal	Filler metal	Brazing temperature (°C)
Aluminium and its alloys	Aluminium-silicon	570-620
Magnesium alloys	Magnesium-aluminium	580-625
Copper and its alloys	Copper-phosphorus	700-925
Ferrous and non-ferrous alloys (except aluminium and magnesium)	Silver and copper alloys, Copper - phosphorus	620-1150
Iron-, nickel-, and cobalt-base alloys	Gold	900-1100
Stainless steels, nickel- and cobalt-base alloys	Nickel - silver	925-1200

Because of diffusion between the filler metal and the base metal mechanical and metallurgical properties of joints can change in subsequent processing or during the service life of brazed components. For example, when titanium is brazed with pure tin filler metal, it is posible for the tin to completely diffuse into the titanium base metal by subsequent aging or heat treatment. When that happens, the joint no longer exists.

6.11.3. Fluxes

Fluxes play a very important part in brazing by (1) dissolving oxides that may be on the surface prior to heating, (2) preventing the formation of oxides during heating, and (3) lowering the surface tension of the molten brazing metal and thus promoting its flow into the joint.

One of the primary factors affecting the quality and uniformity of brazed joints is cleanliness. Although, fluxes will dissolve modest amounts of oxides, they are not cleaners. Before a flux is applied, dirt and oil should be removed from the surfaces that are to be brazed. The less the flux has to do prior to heating, the more effective it will be during the brazing operation.

Borax has been commonly used as a brazing flux. Fused borax should be used because the water in ordinary borax causes bubbling when the flux is heated. Alcohol can be mixed with fused borax when a paste consistency is desired.

A number of modern fluxes are also available that have melting temperatures lower than borax and are somewhat more effective in remvoing oxidation. The particular flux should be selected for compatibility with the base metal being brazed and the particular process being used. Paste fluxes are utilized for furnace, induction, and dip brazing and are usually applied by brushing. Either paste or powdered fluxes can be used with the torch brazing process. Application is usually done by dipping the heated end of the filler wire into the flux.

Fluxes for aluminum are usually mixtures of metallic halide salts, with sodium and potassium chlorides comprising 15% to 85% of the mixture. Activators such as fluorides and lithium compounds are also added. The aluminum brazing fluxes do not dissolve the surface oxide on aluminum.

Since most brazing fluxes are corrosive, the residue should be removed from the work immediately after brazing is completed. This is particularly important in the case of aluminum, to which the chlorides are particularly detrimental. Considerable effort has been directed to developing fluxless procedures for brazing aluminum, as will be discussed later.

6.11.4. Brazing methods

The heating methods used also identify the various brazing processes as described below.

(a) **Torch brazing** : The heat source in torch brazing is oxyfuel gas with a carburizing flame. Brazing is performed by first heating the joint with the torch, then depositing the brazing rod or wire in the joint.

(b) **Furance brazing** : As the name suggests, furnace brazing is carried out in a furnace.

(c) **Induction brazing** : The source of heat in induction brazing is induction heating by high frequency AC current. Parts are preloaded with filler metal and are placed near the induction coils for rapid heating.

(d) **Resistance brazing** : In resistance brazing, the source of heat is through electrical resistance of the components to be brazed.

(e) **Dip brazing** : Dip brazing is carried out by dipping the assembles to be brazed into either a molten filler - metal bath or a molten salt bath (at a temperature just above the melting point of the filler metal), which serves as the heat source. All workpiece surfaces are thus coated with the filler metal.

(f) **Infrared brazing** : The heat source in infrared brazing is a high quartz lamp. The radiant energy is focussed on the joint, and the process can be carried out in vacuum.

(g) **Diffusion brazing** : Diffusion brazing is carried out in a furnance where with proper control of temperature and time the filler metal diffuses into the faying surfaces of the components to be joined.

6.11.5. Braze welding

The joint in braze welding is prepared as in fusion welding. Using an oxyacetylene torch with an oxidizing flame, filler metal is deposited at the joint (Fig. 6.11b), rather than by capillary action as in brazing. Thus considerably more filler metal is used, compared to brazing. However, temperatures in braze welding are generally lower than in fusion welding, and part distortion is minimal. The use of a flux is essential in this process. The principal use of braze welding is to maintain and repair parts, such as ferrous castings and steel components.

6.11.6. Comparison with welding

In comparison with welding, the brazing process is different in a number of ways.

(1) The composition of the brazing alloy is significantly different from the base metal.

(2) The strength of the brazing alloy is sustantially lower than the base metal.

(3) The melting point of the brazing alloy is lower than that of the base metal, so the base metal is not melted.

(4) Bonding requires capillary action.

6.11.7. Advantages of Brazing

Because of these differences state above, the brazing process has several distinct advantages.

(1) Virtually all metals can be joined by some type of brazing metal.

(2) The process is ideally suited for dissimilar metals, such as joining of ferrous to nonferrous, or metals with widely different melting points.

(3) Since less heating is required than for welding, the process can be performed quickly and economically.

(4) The lower temperatures reduce problems associated with heat-affected zones, warping, or distortion. Thinner and more complex assemblies can be joined successfully.

(5) Brazing is highly adaptable to automation and performs well in mass production of delicate assemblies. A strong permanent joint is formed.

6.11.8. Disadvantages of Brazing

The major disadvantage of brazing is that subsequent heating can cause inadvertent melting of the braze metal, weakening or destroying the joint. Too often brazed joints fail when peoply apply heat in an atempt to straighten or repair damaged assemblies. While this is certainly not the result of defective brazing, it can lead to most unfortunate consequences. Therefore, if brazing is used in the manufacture of products that may be subject to service or repair adequate warning should be provided.

6.12. SOLDERING

In soldering, the filler metal, called solder, melts below 450°C. As in brazing, the solder fills the joint by capillary action between closely fitting or closely placed components. Heat sources for soldering are usually soldering irons, torches, or ovens. Soldering with copper-gold and tin-lead alloys was first practiced as long ago as 4000-3000 B.C.

Solders are usually tin-lead alloys in various proportions. For better joint strength and special applications, other solder compositions that can be used are tin-zinc, lead-silver, cadmium- silver, and zinc-aluminium alloys (Table 6.5). Because of the toxicity of lead and its adverse effects on the environment, lead-free solders are being developed and are now in wider use. Typical compositions are 96.5% Sn - 3.5% Ag and 42% Sn - 58% Bi.

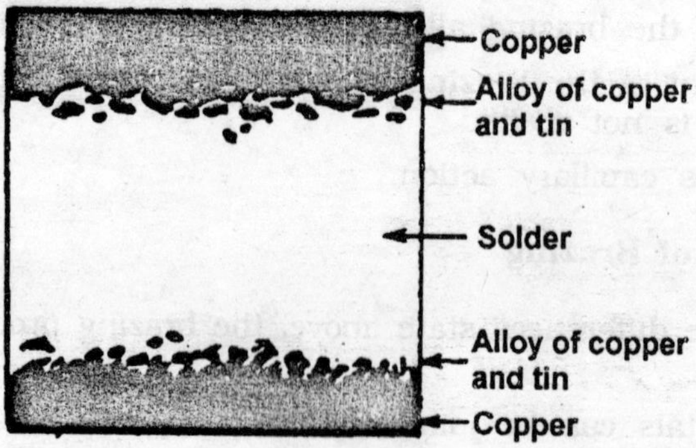

Fig. 6.14. Soldered joint x 300 approx.

Fig. 6.14. illustrate the joint between the two copper surfaces, the layer of metal where the solder joins the copper being an alloy of copper and tin.

Table 6.5. Types of solders and their applications.

Solders	Application
Tin-lead	General purpose
Tin-zinc	Aluminium
Lead-silver	Strength at higher than room temperature
Cadmium - silver	Strength at high temperature
Zinc-aluminium	Aluminum; corrosion resistance
Tin-silver	Electronics
Tin-bismuth	Electronics

In soldering, fluxes are used as in welding and brazing and for the same purposes. Fluxes are generally of two types.

(1) Inorganic acids or salts, such as zinc ammonium chloride solutions, which clean the surfaces rapidly. After soldering, the flux residue should be removed by washing thoroughly with water to avoid corrosion.

2. Non corrosive resin-based fluxes, used in electrical applications.

A good way of heating an iron is by placing it inside an iron or steel box, open at the ends and bottom, and heated inside by the flames from a rectangular gas heater (Fig. 6.15). Electric and gas heated irons may also be obtained and are preferred by some people. Sketches of straight and hotehet-shaped irons are shown in Fig. 6.15, the hatchet type being useful where working with the straight iron is likely to be awkward.

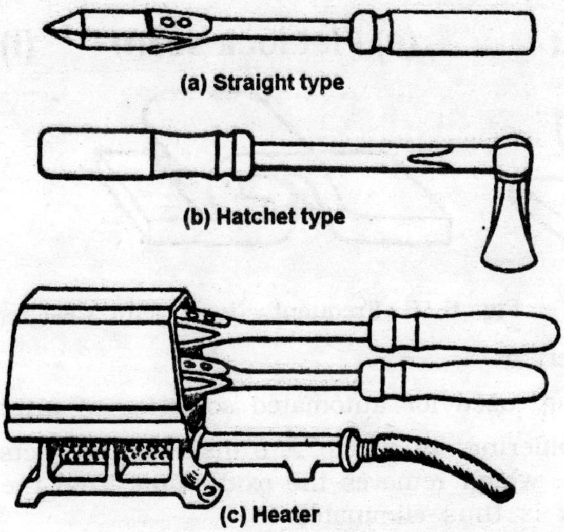

Fig. 6.15. Sketches of straight and hatched shaped irons and heater.

Soldering is used extensively in the electronics industry. Unlike brazing, soldering temperatures are relatively low. Thus a soldered joint has very limited use at elevated temperatures. Moreover, because solders do not generally have much strength, they are not used for load-bearing structural members. Because of the small faying surfaces, butt joints are rarely made with solders. In other situations, joint strength is improved by mechanical interlocking of the joint.

Copper and precious metals, such as silver and gold, are easy to solder. Aluminium and stainless steels are difficult to solder because of their strong, thin oxide film. However, these and other metals can be soldered using special fluxes that modify surfaces. Joint designs commonly used for soldering are shown in Fig. 6.16.

There are several soldering methods, which are similar to brazing methods.

1. Torch soldering
2. Furnance soldering
3. Iron soldering, using a soldering iron.
4. Induction soldering
5. Resistance soldering
6. Dip soldering

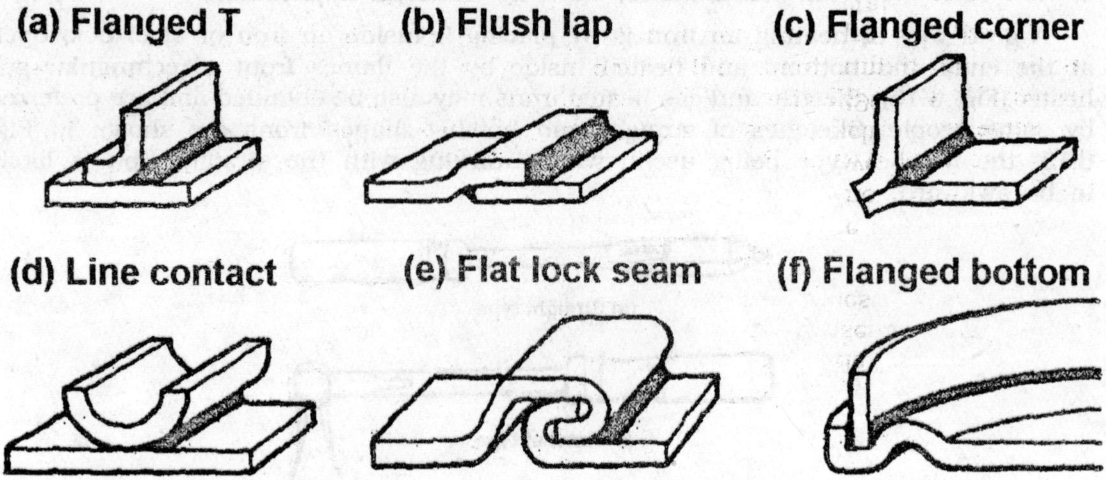

Fig. 6.16. Frequently used joint designs

7. Infrared soldering

8. Wave soldering, used for automated soldering of printed circuit boards.

9. Ultrasonic soldering, in which a transducer subjects the molten solder to ultrasonic cavitation, which removes the oxide films from the surfaces to be joined. The need for a flux is thus eliminated.

EXERCISES

A. Theory

6.1. Define Welding

6.2. Define fusion welding and solid state welding

6.3. Explain what is meant by solid-state welding

6.4 What is meant by welding? How does it differ from soldering and brazing

6.5. List the general weldability, specific metals and specific techniques are used.

6.6. Draw a neat sketch and identify the three distinct zones

6.7. Describe briefly the heat affected zone (HAZ). Draw a neat sketch of HAZ for low carbon steel.

6.8. Draw a schematic diagram of various regions in a fusion weld zone and corresponding phase diagram for 0.3 per cent carbon steel.

6.9. How does the weldability of steel changes as its carbon content increases? Why?

6.10. Discuss the weldability of several metals and explain why some metals are easier to weld than others. Cast iron is generally difficult to weld. Why?

6.11. With the aid of diagrams, show what is meant by?
(i) Carburising flame; (ii) a neutral flame;
(iii) an oxidising flame and explain where each is used.

6.12. Explain with the aid of sketches, how plain carbon steels are flame cut.

6.13. Discuss advantages and limitations of the oxy-acetylene welding.

6.14. Disucss the major hazards associated with oxy-acetylene welding and describe how these can be minimised.

6.15. What do you understand by gas welding? Draw and explain basic welding equipment. Draw a weld zone for oxy-acetylene welding.

6.16. Draw a sketch showing the types of flame available by varying the blow pipe control settings.

6.17. Draw sketches of axy-acetylene cutting equipment and flame cutting machines.

6.18. What do you understand by oxy-acetylene gas cutting? Draw and explain oxy-acetylene cutting equipment and flame cutting machines.

6.19. Explain what is meant by shielded arc welding. Draw a schematic illustration of the process.

6.20. Draw a sketch which is showing the shielded metal arc welding operation.

6.21. Explain briefly the essential differences between MIG and TIG welding.

6.22. Sketch either a TIG welding gun or a MIG welding gun and label its more important features.

6.23. With the aid of sketches describe the submerged - arc welding process and discuss its advantages and limitations.

6.24. Explain what is meant by plasma arc welding. Differentiate between Transferred and Non transferred methods.

6.25. With the aid of sketch describe flux cored arc welding process.

6.26. Explain why the electroslag welding process is suitable for thick plates and heavy structural sections.

6.27. With the aid of sketch describe electrogas welding process.

6.28. Draw a neat sketch of equipment used for electroslag welding operations.

6.29. Write down the major advantages of laser beam welding over Electron beam welding.

6.30. Explain briefly laser beam welding and electron beam welding processes.

6.31. Explain briefly the essential differences between laser beam welding and electron beam welding.

6.32. Why are there problems in welding high carbon steels?

6.33. Explain the reasons for preheating and post-welding heat treatment.

6.34. Describe and explain the hardness and structural changes that can occur with the welding of non-heat-treatable and heat-treatable aluminium alloys.

6.35. Compare the advantages and limitations of the following fusion welding processes.
(a) Oxy-acetyle welding (b) Metallic arc welding
(c) TIG welding (d) MIG welding (e) Submerged arc welding.

6.36. Select suitable welding processes for the following applications, giving reasons for your choice.
(a) Assembly of sheet steel components under productin conditions.
(b) The repair of motor car body panels, (c) The fabrication of steel girders on site; (d) The assembly of airframe components in aluminium alloy.
(e) The fabrication of large power-press frames from thick steel plate.

6.37. Explain the factors involved in electrode selection in arc- welding processes.

6.38. What are the similarities and differences between consumable and nonconsumable electrode arc welding processes.

6.39. With the aid of sketches, show the effect of welding on the crystal structure of : (a) mild steel (b) copper (c) aluminium.

6.40. Do you think it is acceptable to differentiate brazing and soldering arbitrarily by temperature of application? Comment.

6.41. State and explain the factors that determine the strength of a brazed joint.

6.42. Explain the effect of brazing on structure and properties in a brazed joint.

6.43. Distinguish between soldering, brazing, fusion welding and cold welding.

6.44. Differentiate between soldering, brazing and welding processes. Also write their specific fields of applications.

6.45. Differentiate between brazing and soldering. Describe any one of these process with applications.

6.46. Differentiate between brazing and braze welding.

6.47. List the filler metals which are used in brazing.

6.48. List the typical filler metal, base metal and their corresponding brazing temperature in °C.

6.49. Name the different types of brazing methods and explain any two.

6.50. What are the uses of flux in brazing.

6.51. What are the uses of filler metals in brazing.

6.52. Name the different types of solders and their corresponding applications.

6.53. Name the two types of fluxes used in soldering.

6.54. Why we are using lead free solders? Give the typical compositions.

6.55. Name the different types of soldering methods. Expian briefly wave and ultrasonic soldering.

6.56. Where would you go for gas welding. Arc welding soldering and brazing? **[MREC 2001]**

6.57. Write short notes on "Arc Welding". **[MREC 2001]**

6.58. What is meant by welding? How does it differ from soldering and brazing? Compare the effects of straight and reverse polarity. [RU 2001]

6.59. Write short notes on "left hand & right hand methods of welding". [MREC 2000]

6.60. Write short notes on "Brazing and soldering". [RU 2000]

6.61. What do you understand by term 'polarity'? How is an electrode specified? What factors govern the selection of an electrode? [RU 1999]

6.62. Write short notes on
(i) Gas welding flames (ii) Braze welding [RU 1999]

6.63. Explain the principle of arc welding. List the equipments and tools used in it. [RU 1998]

6.64. Write short notes on (i) Oxyacetylene gas welding (ii) Resistance welding [RU 1998]

6.65. (a) Describe the principle of Oxy-Acetylene gas-cutting. Differentiate between gas-cutting torch and welding torch [RU 1997]

6.66. (a) Give a general classification of welding Operations, (b) Distinguish between brazing and soldering. What are their applications? [RU 1996]

6.67. Draw a tree diagram showing general classification of welding process. [RU 1995]

INDEX

Accelerator strain aging 46
Age hardening, 95, 100
Alloy steels 74
Aluminium alloys 98
Aluminium bronzes 104
Annealing 84, 88
Annealing twins 29
Arc welding 181
Arsenical copper 103
Atomic radius 18
Austempered ductile irons 82, 83
Austempering 95
Austenite 63
Austenitic 79
Babbitt alloys 109
Backhand welding 181
Bainite 65
Bearing material 109
Bending (cold working) 143
Bending 129
Beryllium 104
Black heart 83
Blow pipe 189
Blue brittleness 46
Brasses 104
Braze welding 193
Brazing 188
Brazing hearth 189
Brazing metals 189
Brazing methods 192
Brittleness 9
Bulk defects 36
Burger circuits 35
Carbonitriding 93
Case hardening 93
Cast irons 80
C-curve 71
Cementite 63
Characteristics of crystal structure 19
Charpy test 58
Chisels 123
Clevage planes 42
Coining 146
Cold forging 141
Cold heading 141
Cold working 139
Commercial annealing 90
Cooling characteristics of casting 172
Cope and drag pattern 165
Copper alloys 103
Copper magnesium zinc alloy 98
Core and core prints 170
Corrosion resistant steels 78
Creep 52
Crystal directions and planes 22
Crystal modes 14
Crystallization 12
Cupronickel 104
Cyaniding 94
Deep drawing 134
Defects 46
Defects in solid crystals 30
Deformation of materials 24
Deformation of slip 26
Delta-iron 65
Density of crystal 19
Diffusion brazing 193
Dip brazing 193

Soldering 194
Solid state welding 175
Sorbite 66
Spheroidizing annealing 90
Spinning 134, 144
Split pattern 164
Squeezing 145
Stainless steels 78
Straight and hatched type iron 195
Strain aging 45
Strain twins 29
Subcritical annealing 90
Surface hardening 93
Swages 124
Swaging 120
Sweep pattern 167
Tempering 84, 92
Tensile test 4
Teritary creep 53
Tin bronzes 104
Tongs 127
Torch brazing 192
Tough pitch copper 103
Toughness 10
Transition temperature 44
Troostite 66
T-T-T diagrams 71
Tube drawing 142
Twinning 27
Ultimate tensile strength 6
Upsetting 120, 127
Void formation 44
Wear resisting steels 77
Welding 175
White cast irons 80
White heart 83
White iron 82
Work hardening 37
Wrought material 98
Yield point 5
Young's modulus 9

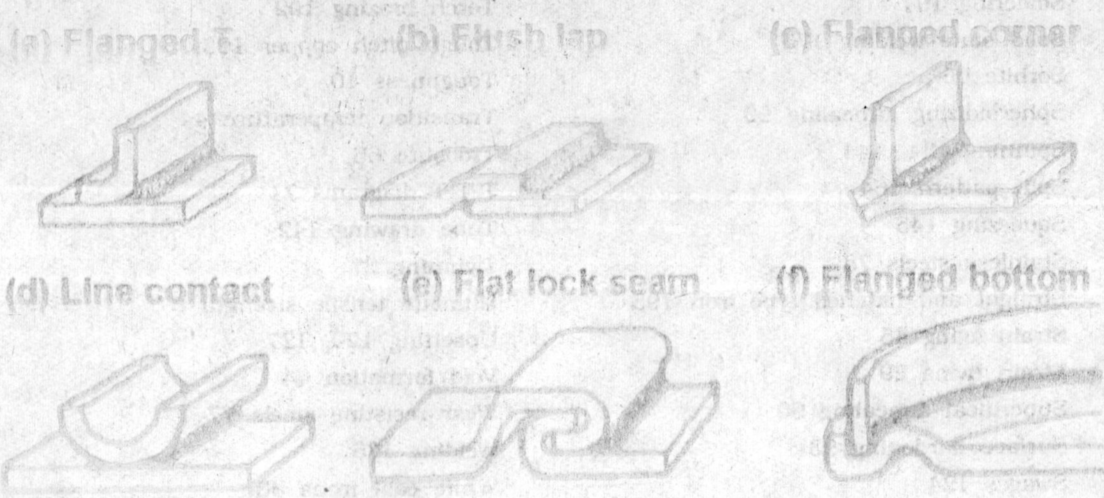

Fig. 6.16. Frequently used joint designs

7. Infrared soldering.

8. Wave soldering for automated soldering of printed circuit boards.

9. Ultrasonic soldering in which a transducer subjects the molten solder to ultrasonic cavitations which remove the oxide films from the surfaces to be joined. The need for a flux is thus eliminated.

EXERCISES

A. Theory

6.1. Define Welding.

6.2. Define fusion welding and solid state welding.

6.3. Explain what is meant by solid state welding.

6.4. What is meant by welding? How does it differ from soldering and brazing?

6.5. List the general weldability, specific metals and specific techniques are used.

6.6. Draw a neat sketch and identify the three distinct zones.

6.7. Describe briefly the heat-affected zone (HAZ). Draw a neat sketch of HAZ for low-carbon steel.

6.8. Draw a schematic diagram of various regions in a fusion weld zone and corresponding phase diagram for 0.3 per cent carbon steel.

6.9. How does the weldability of steel changes as its carbon content increases? Why?

6.10. Discuss the weldability of several metals and explain why some metals are easier to weld than others. Cast iron is generally difficult to weld. Why?